FOREWORD

This study presents an international review of cost estimates for decommissioning of nuclear facilities. The objective of the study is to illustrate reasons why the published estimates for decommissioning vary so widely and to see to what extent various political, institutional, technical and economical factors could explain the variation. The report is intended for the general reader with an interest in the topic.

The work has been carried out by an international group of experts under the auspicies of the Nuclear Energy Agency's Committee for Technical and Economic Studies on Nuclear Energy Develoment and the Fuel Cycle (NDC). The report does not necessarily represent the views of Member governments or participating organisations. The report is published on the responsibility of the Secretary-General of the OECD.

TABLE OF CONTENTS

List of Tables

List of Figures

ANNEXES

EXECUTIVE SUMMARY

Several international and national studies have shown that technical methods and equipment are available today to safely dismantle nuclear facilities, of whatever type or size. Considerable experience already exists in the use of these techniques in maintenance and repair work, and in decommissioning of prototype, demonstration, and small power reactors. Several projects to decommission various types of nuclear facilities are also under way. For the currently operating commercial nuclear power plants and fuel cycle facilities, it can be concluded that the decommissioning costs will be affordable, adding only a few per cent to the total generating cost of nuclear electricity.

However, the individual cost estimates show relatively large variations. The goal of this study is to identify the reasons for these variations, and to consider to what extent various political, institutional, technical and economic factors can explain the differences. The focus is on decommissioning cost estimates for commercial nuclear power plants.

Cost assessment methods

In general, the basis of the cost estimates for decommissioning projects lies in the worldwide experience so far obtained either in decommissioning projects or in maintenance and repair work at operating nuclear facilities where conditions are similar. This experience can be utilised directly or as an analogue for estimating the costs of similar tasks in decommissioning projects, or indirectly for the assessment of unit costs for basic decontamination and dismantling activities.

Different costing methods have different data requirements, and consequently, their reliability depends on the extent to which various data are available and applicable to the specific case being considered. Independent of the assessment method, some uncertainty is inevitable in all estimates of future costs, and no costing method is generally superior to others in this respect. However, analysis of the costing method may be useful in order to locate the key uncertainties in each specific estimate.

Reasons for differences in cost estimates

A general difficulty in international cost comparisons is the ambiguity of monetary conversions. In particular, the wide variations in the exchange rates between the US dollar and other currencies through the 1980s make the comparisons sensitive to the base year of the monetary conversions. The inflation rates, which may affect the exchange rates, also have to be taken into account when comparing estimates from different years.

Furthermore, a distinction should always be made between discounted and undiscounted costs. In the cases where decommissioning work extends over many decades, even a moderately low discounting rate makes the present-worth of the total costs substantially smaller than the basic undiscounted costs. The discounted cost estimates are useful for financing considerations and for comparison of various decommissioning strategies for the same facility, but for understanding the differences between estimates for different plants, undiscounted cost estimates are more appropriate.

However, a considerable part of the differences between decommissioning cost estimates are independent of the monetary conversion factor. They stem from the physical differences between the facilities themselves, and from the assumptions underlying the decommissioning plans and cost estimates. Several factors explain why these assumptions are different in different estimates.

National policy decisions and the general legal and institutional framework that control the use of nuclear energy in individual countries also constrain the overall strategies on which future decommissioning plans are based. Various regulatory rules and principles must be taken into account in the planning of the decommissioning work.

Because of differences in these legal and regulatory policy frameworks, the decommissioning plans will be different both in the scope of work and in the timing and implementation of the individual activities. According to some plans, all facility components and buildings are to be removed and the site returned to a greenfield condition within a few years of the shutdown of the facility. Other plans only call for the removal of the radioactive parts of the plant, which is required for licence termination. In some cases, the dismantling may be projected to start several decades after the shutdown, allowing the radioactivity to decay to lower levels. All these variations in the plans lead to differences in cost estimates as well.

Another major reason for variation in the plans and cost estimates is the differences between the facilities themselves. The extent of effort needed in different decommissioning projects depends on the amount of materials and equipment that need to be removed. In this respect, both the type and the specific design of the reactor are important. The power capacity of the plant affects the physical dimensions of the reactor pressure vessel and other major components, but its effect may be easily overridden in importance by the differences in other technical characteristics. Therefore, the comparison of decommissioning cost estimates on the basis of costs per capacity unit is misleading.

The differences in decommissioning plans and physical characteristics of the facilities are reflected in the estimated amount of manpower needed in the decommissioning work and in the amount of waste arisings. In spite of the apparently large discrepancies between the cost estimates submitted for the study, the estimated costs per unit weight of radioactive waste produced turned out to be relatively close to each other. Account must be taken of the differences in scope of these cost estimates: for instance, because of the unresolved situation in waste management and disposal, these costs have so far been excluded from some estimates.

Finally, the unit costs of labour and services vary by country and region, and different assumptions are made regarding their future evolution. Different assumptions can also be made concerning the costs of future waste management and disposal, as practical disposal solutions may still be pending final decision by authorities. The contingency allowances that are made for uncertainties in technical plans and future cost development, depend on the degree of conservatism required, which will, in turn, be linked to the purpose of the estimate.

Conclusion

Because of the significant differences in several factors underlying the estimates, there is no reason to expect that the decommissioning cost estimates should be similar or even nearly similar for separate plants. Each plant is different, as is each decommissioning plan. The range of variation between cost estimates is significantly narrowed when the estimated costs are proportioned to the estimated amounts of waste arisings. The remaining variation can be largely explained by differences in technical details of the facilities and by different unit costs for labour and services. Furthermore, any international cost comparison is subject to the ambiguities arising from the monetary conversions.

Decommissioning cost estimates are subject to uncertainties, and these uncertainties are treated in different ways in different estimates. However, a vast amount of technical experience has been accumulated on which the judgments on engineering costs can be based. Ongoing work in several OECD countries will further enhance this data base. Information from a number of completed decommissioning projects already gives evidence that the decommissioning costs can be estimated with reasonable accuracy.

1. INTRODUCTION

1.1 Goal and scope

This study presents an international review of cost estimates for decommissioning of nuclear facilities. The objective of the study is to examine reasons why the published estimates for decommissioning vary so widely and to see to what extent various political, institutional, technical and economic factors can explain the variation. The principal effort has been concentrated on identifying the factors that appear important for the understanding of cost estimates for the future decommissioning of commercial nuclear power plants. Some information on decommissioning cost estimates for nuclear fuel cycle facilities is provided in Annex 1.

The report is intended for a general reader with an interest in the topic. Chapter 2 defines the three stages that are commonly used to describe different decommissioning strategies. Costing methods are briefly discussed in Chapter 3 and examples of present cost estimates from OECD countries are introduced in Chapter 4. These estimates form the main reference for the subsequent discussion of cost factors in Chapter 5.

The study has been undertaken under the auspices of the Nuclear Energy Agency's (NEA's) Committee for Technical and Economic Studies on Nuclear Energy Development and the Fuel Cycle. An Expert Group with members from 11 countries and two international organisations was established for the study.

1.2 Background

In 1986 the NEA published an Expert Group report on decommissioning of nuclear facilities, which concluded that "the decommissioning of commercial reactors and supporting fuel cycle facilities is considered to be technically feasible, the waste volumes manageable and the costs affordable" [1]. On the basis of the cost estimates included in the report, it was shown that decommissioning adds very little to the costs of nuclear electricity as its share in no case considered was larger than a few per cent of the total generating cost.

The cost conclusion was reconfirmed by the more recent report by the NEA and the International Energy Agency on the costs of electricity generation [2]. In this report, for which information was gathered from electricity utilities and some governmental administrations, the judgement was made that decommissioning normally adds less than one quarter of a per cent to the total levelized generating costs. Even a doubling of the estimate of decommissioning costs only led to a contribution of 1.1 per cent to generation costs in the highest example reported.

However, large variations in the published estimates for decommissioning costs have raised questions regarding the reliability of present estimates. A recent comparison of decommissioning cost estimates in the USA showed a variation by a factor of 4 between the estimates [3]. These estimates were for 67 commercial light water reactor plants of different sizes and ages, and one high temperature gas cooled reactor. In the estimates submitted for the present study from nine countries within the OECD, the variation is considerably larger, but so is also the variety of reactor types, sizes and ages considered.

10

The doubts about the decommissioning cost estimates have been said to imply that the present provisions made for the future decommissioning activities may be insufficient. In several countries systems have been established to collect a fund during the reactor's operating period which, possibly together with the interest accrued, would be enough at the time of decommissioning to cover the costs. The size of the annual contributions to the fund is usually based on the present cost estimates. In some countries, the power plant owner's financial responsibility for decommissioning is absolute: in case the funds collected for decommissioning eventually turn out to be insufficient, the owners have to pay the outstanding expenses from their current revenues. Therefore, general equity considerations may require that the cost estimates should correspond to the eventual costs as closely as possible.

Uncertainties are inevitable in all economic assessments of future projects. However, there is no longer uncertainty about technical feasibility of decommissioning. Several pilot projects and international studies have shown that technical means and equipment already exist today to carry out a decommissioning project, even for large commercial power plants (e.g. [4] and [1]). Extensive experience has been obtained in using these methods for repair and maintenance work at operating nuclear facilities. Therefore, a considerable amount of expertise already exists on which to base the estimates of the decommissioning costs.

One obvious reason for variation in the cost estimates is the variation in facility characteristics. A quick look at utilities' decommissioning plans reveals further reasons for variations: the scope of the work to be done at decommissioning varies considerably as do the plans for final disposal of the wastes. In principle, it would be of interest for the sake of comparison to try and adopt a common reference definition for decommissioning and subsequent waste disposal and to prepare the cost estimates for such "standardized" decommissioning plans. The remaining variation could then be attributed to differences in facility details, economic conditions and experts' judgements. However, such a standardized plan would be artificial, since the legal and regulatory frameworks are, indeed, different in different countries. Therefore, instead of such an artificial comparison, an attempt has been made to identify the factors underlying the differences and to put these factors in perspective as to their relative importance.

1.3 Other NEA work on decommissioning

Several of the completed and current decommissioning projects in OECD countries are the subject of the NEA's "Co-operative Programme for the Exchange of Scientific and Technical Information Concerning Nuclear Installation Decommissioning Projects" [5]. The Programme is described in outline in Annex 5. A Task Group of the Programme is trying to identify the cause of differences in the cost estimates of the projects under the Programme while the present report focuses on the cost estimates for future decommissioning of commercial nuclear power plants. The Task Group has worked on a unified terminology, forming a cross-comparison of twelve of the projects in the Programme on the basis of the cost structures of these projects. The present report has greatly benefited from a close interchange with the work of that Programme and preparation of it has been facilitated by overlapping membership of some of the experts in the Expert Group and in the Programme's Task Group on decommissioning cost.

1.4 Sister programme on final disposal costs

Parallel to this work, a study has been done on costs of final disposal in geological repositories. The objective of this sister project is, like the present study, to illustrate the origins of variations in disposal cost estimates of spent nuclear fuel, or high level vitrified waste and long-lived alpha bearing waste. The study started in October 1990 under the auspices of the NEA and its report will be published in 1992.

2. DECOMMISSIONING STRATEGIES AND ALTERNATIVES

The decision on how to perform the decommissioning of a particular nuclear facility is based on the consideration of a number of factors. These are likely to include national nuclear strategy, facility characteristics, regulations, alternative options for radioactive waste management and disposal, costs, and social effects. As each of these factors may have widely different meanings and implications in any particular circumstance, the strategies for carrying out the practical decommissioning work may vary largely, both in technical detail and in schedule.

The three principle "stages" of decommissioning are often used to describe decommissioning strategies [6]. Each stage corresponds to a certain physical state of the plant and its equipment, and is also characterised by the surveillance requirements necessitated by the states. The main features of these stages are illustrated in Figure 1 for nuclear power plants in a case where decommissioning includes dismantling of all buildings and equipment. The stages are applicable to most nuclear fuel cycle facilities such as fuel fabrication plants and reprocessing plants as well, but they are not especially suitable for uranium mines or waste repositories.

A practical decommissioning strategy may comprise one stage only or proceed through several stages. Stage-1 and 2 are sometimes considered as alternatives to each other. Although in most countries these two stages are considered only as interim steps on the way to eventual Stage-3 or something like it, some of the plans have so far left open how and when the final stage of decommissioning should be reached.

The following definitions of the three stages are descriptive and they may not be directly applicable for any practical decommissioning plan. Their principal intent is to facilitate discussion of different decommissioning approaches.

In most cases the nuclear fuel and unattached radioactive materials in the process systems, as well as radioactive waste produced in normal operation, will have been removed before decommissioning proper begins. In some cases, however, these activities are treated as part of the decommissioning process and their costs are included in the cost estimates.

2.1 Stage-1 decommissioning

State of the plant and equipment

For reactors, the first contamination barrier is kept as it was during operation but the mechanical opening systems (valves, plugs, etc.) are blocked and sealed. The containment building is kept in a state appropriate to the remaining hazard. The atmosphere inside the containment building is subject to appropriate control of composition such as humidity and radioactivity. Access to the inside of the containment building is controlled, and personnel entering and exiting the containment are monitored for radioactivity. For fuel cycle facilities some of the mechanical operating systems may be retained for use during decontamination work during Stage-2.

Surveillance

The plant is kept under surveillance with routine inspections to assure that the plant remains in good condition. The equipment necessary for monitoring radioactivity both inside and in the area around the plant is kept in good condition and used when necessary and in accordance with national legal requirements. Checks are carried out to ensure that there are no leaks in the first contamination barrier nor in the containment building.

2.2   Stage-2 decommissioning

State of plant and equipment

For reactors, the first contamination barrier is reduced to its minimum size (all parts easily dismantled are removed and the remaining barrier is sealed). The sealing of the barrier is reinforced by physical means and the biological shield is extended, if necessary, to surround the barrier completely. After decontamination to acceptable levels, the containment building and the ventilation system may be modified or removed if they no longer play a role in radiological safety. Depending on the extent to which other equipment is removed or decontaminated, access to the containment building, if it has been left standing, can be permitted. The non-radioactive parts of the plant (buildings or equipment) may be converted for new purposes. For fuel cycle facilities, the primary radioactive plant and equipment will sometimes be removed during Stage-2.

Surveillance

Surveillance around the barrier may be reduced but it is desirable for periodic spot checks as well as surveillance of the environment to be continued. External inspection of the sealed parts is performed. Measurements for leaks are no longer necessary on the remaining containment building.

2.3   Stage-3 decommissioning

State of plant and equipment

Any materials, equipment and parts of the plant whose activity levels remain significantly above natural background despite decontamination procedures, are removed. In all the remaining parts, contamination is below the authorized release limit, a very low level approaching that of the natural environment except in those cases where a higher level can be accepted as the site is to be re-used for other nuclear purposes.

Surveillance

Unless re-used, the site is released without any access restrictions arising from the residual radioactivity. From the point of view of radiological protection, no further surveillance is necessary.

3. COSTING METHODS FOR DECOMMISSIONING ACTIVITIES

3.1 Basis of cost estimation

The decision on decommissioning strategy requires that some picture exists also on the costs of the alternatives, and, hence, that some basic principles have been fixed on how the work would be done in alternative strategies. Once the decision has been made, a detailed technical plan can be prepared for the practical work. Such a plan can then be used as a basis for detailed cost estimation.

The assessment of the costs for work activities, materials and services can be based on previous experience from similar projects or from other technical fields. The past and current decommissioning projects, such as those described in Annex 2 offer valuable data. Much of the experience on repair and maintenance work at nuclear facilities is also directly applicable for decommissioning work and can be used to estimate work efficiencies — the application of some working methods may, in fact, be more straight-forward in decommissioning as no special care has to be taken to avoid damage to structures. The efficiency data from other technical fields may have to be modified to account for the specific circumstances, e.g. the necessity to protect against radiation.

Site-specific unit cost data will be needed to complete the cost estimation. Since decommissioning is, in general, a labour-intensive activity, assumptions on labour costs will be particularly important. Uncertainties will probably be inevitable in any cost assessment, but their meaning may be different in different contexts. Hence, the magnitude of contingency allowances that are added to the basic cost estimate reflects both the extent of residual uncertainties and the degree of conservatism that is sought.

3.2 Costing principles

Practical costing of decommissioning projects requires the identification of all work activities together with their associated material, equipment and service requirements, and the subsequent estimation of the costs arising from each activity. The level of itemization reflects the costing approach adopted. In some cases, the work on each separated main component, system or room at the plant is costed separately whereas in some other cases each activity is split to a series of discrete and measurable elementary work activities for which the unit prices or costs will be determined.

In the former case the emphasis is on costing experience from other projects with similar characteristics. For some work items such as removal of pumps or boilers cost data may already exist which can be used directly — or perhaps after a modification to account for plant specific features. Cost estimation for construction projects is similar to estimation for decommissioning in many aspects. For work items for which experience is still scarce, the estimate has to be produced starting from phase by phase review of the effort needed. Engineering judgment is needed to assess manpower requirements, work efficiencies and time schedules.

In the latter case basic cost estimates are developed for elementary repetitive activities such as cutting a unit length of pipe or removing a valve, a pump, or a unit quantity of concrete, etc. Various coefficients are used to allow for specific circumstances. The total cost estimate, hence, become dependent on a number of unit cost factors, which are applied throughout the facility. Here also the cost estimation will rely on experience and engineering judgment but the focus of judgment is on smaller details than in the previous approach. This implies that a much more detailed plant inventory will also be needed.

In practical costing, both approaches can be mixed: unit cost factors are used for those work items which can be easily split down to basic cost components; for other items, specific estimates will be produced separately.

The following text will illustrate various steps of decommissioning cost estimation. The presentation is structured according to the unit cost factor approach, but similar considerations will form the framework of any cost estimate.

3.3 Preparing a total cost estimate

3.3.1 Cost categories

The present discussion of decommissioning cost estimation draws on the guidelines presented in the US National Environmental Studies Project report [7]. Accordingly, the decommissioning costs are classified into three categories: (1) activity-dependent costs, (2) period-dependent costs and (3) collateral costs and special item costs. Each separate cost item may include a contingency allowance.

Activity-dependent costs are directly related to the extent of "hands-on" work at decommissioning and include activities such as decontamination, removal of components and packaging, shipping and disposal of waste. The costs arise from labour, materials, energy, equipment and services.

Period-dependent costs are proportional to the duration of individual tasks or of the whole project. They arise from project management, administration, routine maintenance, radiological, environmental and industrial safety and security. They are often fairly independent of the precise level of the concurrent ongoing hands-on activities.

Collateral costs and special item costs are those that are neither assignable to a certain work activity nor period-dependent. The purchase or rent of equipment may belong to this category if the equipment is used to support many distinct activities.

3.3.2 Identification of activities and inventories

Once the principles are clear, a technical plan can be worked out for implementation of the decommissioning strategy. The planning assumptions have to be listed and the scope of the project must be defined at the operational level. A technical plan is prepared describing how each part of the facility will be brought to the state required to meet the overall goal of the decommissioning strategy.

If the unit cost factor approach is followed, the technical plan must be stated in terms of the discrete basic activities for which the unit costs can be developed. The list of activities must be accompanied by an inventory of the plant buildings and equipment which can be used to determine the extent of each activity. Such an inventory specifies, for instance, the total quantity of piping, classified by size and radioactivity, the number of pumps, valves, heat exchangers, each classified by size and contamination, the amount of concrete, classified on the basis of the size and density of reinforcing steel, etc. Further classification needs may arise from the regulations and criteria governing the waste disposal practices.

3.3.3   Unit cost factors

The development of unit cost factors must accord with the level of itemization in the plant inventory and activity listing. As regards the items mentioned above, unit cost factors would be needed for cutting a unit length of pipe of a certain size and level of contamination, removing one pump of each size and contamination class, etc. One approach is to develop the basic unit cost factors for ideal conditions, e.g. cutting noncontaminated pipe at a worker's waist height without risk of radiation exposure, and then separately to assess the work difficulty factors that are needed to account for the specific work conditions.

The basic unit cost factor incorporates the labour hour requirement per unit activity (under ideal conditions) and the local labour expenses per work hour, taking into account the different worker and craft categories. The difficulty factors may allow for considerations of, e.g. working height, need for protective equipment (respirators, protective clothing), radiation levels (ALARA principle), work breaks and other productivity losses. The final unit cost factor may, additionally, be modified to include the materials and equipment costs whenever these are directly proportional to the extent of work. For waste disposal, the unit cost factors may be given directly per unit volume of packaged waste of each type.

3.3.4   Project schedule and staff requirements

The duration of individual work phases can be calculated on the basis of the plant inventory and labour efficiency data for each basic work activity. The overall project duration will be determined by those activities that are on the critical path; the activity is said to be on the critical path if the startup or continuation of all other remaining tasks eventually depend on its completion. In this way, a time schedule may be produced for different phases of the decommissioning work and for the whole project. This schedule serves as one basis for estimating the period-dependent costs of decommissioning.

The other need is an estimate of the size of the staff that is required for management, administration and other supporting functions. A part of the staff requirements is proportional to the level of actual hands-on activity, but a certain number of personnel will be needed almost independently of the extent of ongoing work. The relation between period-dependent costs and activity-dependent costs may give rise to a need for optimization, since the project duration can often be reduced by increasing the crew size in the activities on the critical path. However, the owner's staff requirements may be small at the actual stage of decommissioning, if most of the decommissioning activities are split into well-defined work packages that are to be carried out by outside contractors.

3.3.5   Collateral and special item costs

The costs of such items as:

- heavy equipment for site support,
- small tools,
- health physics equipment and supplies,
- licenses and permits,
- nuclear liability insurances,

may be dependent neither on the level of activities, nor on the duration of the project. They may, therefore, be considered as a separate cost category. Energy costs (lighting, heating, cooling) may also be included in this class, although at least a part of the energy requirement is proportional to the duration of some project phases.

For some parts of the equipment removed, a scrap or salvage value could be contemplated. However, it seems to be uncertain if such value can ever be materialized, therefore these considerations are often ignored in the final cost estimates.

3.3.6 Total costs

The total cost estimate is obtained as a sum of the costs in the three categories. Basically, the activity-dependent costs will be calculated on the basis of activity lists, plant inventories and unit cost factors; the period-dependent costs on the basis of estimates, project schedules and staff requirements; the collateral costs will be assessed separately for each item. However, before summing up, the cost estimates may be adjusted to include a contingency that reflects the uncertainty in the estimates. Alternatively, or sometimes additionally, a common contingency factor may be applied to the total cost estimate.

4. COST ESTIMATES FOR DECOMMISSIONING OF NUCLEAR POWER PLANTS

4.1 Cost estimates provided for this study

Results of the questionnaire

For the purpose of this study a questionnaire was circulated to the countries represented in the Expert Group. They were requested to provide information on recent cost estimates for future decommissioning of commercial nuclear power facilities. Actual cost estimates were obtained from nine OECD countries. They covered nuclear power plants of various types, sizes and ages, and were accompanied by explanations on their underlying assumptions. In addition, some cost information was received on decommissioning nuclear fuel cycle facilities.

The cost estimates for nuclear power plants are listed in Table 1. The cost estimates are arranged by the reactor type. To facilitate later discussion in this report a short hand designation is attached to each estimate in the third column of the table.

The power capacity shown corresponds to the design gross capacity (references to power capacities later in this report will also mean gross capacities). The column titled "Mode of Decommissioning" gives a simplified description of the decommissioning strategy that has been used as a basis for cost estimation. Here, simple "Stage-3" basically means a strategy where dismantling of the plant is to start immediately or very shortly after plant shutdown and continue without delays or intermissions up to the final goal set for decommissioning. Combination of several stages and a year-number indicates a stagewise strategy where partial decontamination and dismantling campaigns are followed by storage periods of different lengths until the final goal is reached. The number of years indicates the length of the storage period.

The cost estimates are first shown in the original form in which the Expert Group received them. The number in parenthesis after the currency unit indicates the base year for the money value. It may be different from the year of original estimate (shown in a separate column), in which case the correction for inflation and relative cost escalation has already been done in the estimate submitted.

The conversion to US dollars has been done by the NEA Secretariat. First, the estimates have been changed to correspond to the same base year and month, which was chosen to be January 1990. For this conversion, use was made of the Consumer Price Index (CPI) ratio between January of the year of estimate and January 1990 as shown in the table. Secondly, the modified estimates have been converted to US dollars using the actual exchange rates between dollars and other national currencies as of January 1990. Some unavoidable problems and biases involved in the conversion will be discussed in the next chapter.

Cost information was also obtained for decommissioning of spent fuel reprocessing facilities. Both on-going decommissioning projects and plans for future activities were represented. A summary of these estimates is included in Annex 1.

Apart from uranium mines and mills, reprocessing plants are likely to be the fuel cycle facilities requiring the biggest decommissioning effort. At other facilities, such as fuel fabrication facilities or enrichment facilities, the amounts of radioactive wastes are usually relatively small and the radiation levels low. Therefore, the cases included in Annex 1 can also be considered as examples of the level of costs that decommissioning of fuel cycle facilities may lead to.

Much of the discussion of the cost factors in Chapter 5 is relevant also to fuel enrichment, fabrication and reprocessing facilities. The working methods that are planned to be used at power plants can also be used at fuel cycle facilities. However, the structures and equipment at these facilities are quite different from those at power plants and direct cost comparisons, therefore, are hardly useful.

The main issue for decommissioning of uranium mines and mills is usually the final disposal of mill tailings that will have to be addressed at that time. Hence the problems arising are usually quite different from those at power plants or other fuel cycle facilities. An NEA Group of Experts has discussed the mill tailing issues in a report published in 1984 [8].

Background of the estimates for power plants

The cost information submitted to the Expert Group originates from various studies made in the member countries during the last few years. In some cases slight modifications have been made to match the original estimates to the framework of this study. The following presentation is described in alphabetical order by country.

The estimate from Canada (CAN) has been prepared by the Atomic Energy of Canada Ltd. (AECL). The estimate does not address any specific power plant but a reference plant of standard CANDU 600 MWe pressurized heavy-water reactor (HWR). The plan for decommissioning assumes that the plant is first brought to a "static state" (about the same as Stage-1), then stored under minimal surveillance for 32 years, and finally dismantled to Stage-3 (including building demolition and site restoration). It is assumed that the reference plant is located at river- or sea-side and has no cooling tower.

The costs for reactor final shutdown and inspection, defuelling and storage of fuel (approximately 6 years), removal and storage of heavy-water, removal and disposal of replacement pressure tubes[1], and disposal of radioactive wastes from 30 years of operation are included in the decommissioning cost estimate.

The Finnish estimates cover both operating nuclear power stations in the country. The first estimate (FI-1), provided by Imatran Voima Oy (IVO), is for the Loviisa 2 x 465 MWe pressurized water reactor (PWR) station based on the Soviet VVER design but equipped with a Westinghouse type containment with ice condenser. The decommissioning plan assumes complete dismantling of all radioactive components and parts immediately after shut-down of the reactors. Dismantling of non-radioactive parts and buildings is not envisaged in current plans as the site will continue to be used for energy production purposes.

The second estimate from Finland (FI-2) is for the Olkiluoto 2 x 735 MWe boiling water reactor (BWR) station based on Asea Atom's (nowadays Asea Brown Boveri) design with pressure suppression containment and internal main recirculation pumps. The estimate has been provided by Teollisuuden Voima Oy (TVO). A delayed dismantling after a 30 year storage is assumed in the plans, and the dismantling is restricted to the radioactive part only. Both Finnish nuclear power stations are located beside the sea.

The cost estimates are based on recent technical plans for decommissioning and serve as a basis for the provisions that are made for the future decommissioning activities.

The German estimates (GER-1 and GER-2), provided by a group of engineering companies, are for the same facility, Biblis A reactor, a 1204 MWe KWU design PWR. The site has two reactor units but the

---

1. The pressure tubes are assumed to be replaced after 15 years of operation. A 600 MWe CANDU reactor has around 380 pressure tubes (fuel channels).

estimates were based on the independent decommissioning of one of the reactors. The plant is located on the banks of the Rhine river and is equipped with a cooling tower. An immediate dismantling strategy is assumed in GER-1, whereas a delayed dismantling after a 30 year storage in GER-2. In both estimates it is assumed that the site will be brought to a greenfield condition.

The Italian estimate, provided by the Ente Nazionale per l'Energia Elettrica (ENEL) is for 160 MWe dual cycle BWR plant. So far the estimate covers only the costs of bringing the plant to a safe storage condition (Stage-1).

The Japanese estimates are based on a study prepared by the Sub-committee on Nuclear Energy of the Advisory Committee for Energy in 1984. Estimates were submitted both for PWR and for BWR. They do not address any specific power plant but reference plants of 1100 MWe BWR and 1160 MWe PWR, that are considered representative for all large light-water reactors currently used in Japan. All Japanese nuclear power stations are located beside the sea.

The current strategy for decommissioning assumes a short, 5 to 10 year storage period in Stage-1 before final dismantling of the plant. The site is assumed to be returned to a greenfield condition. Costs for radioactive waste disposal are not included, pending the formulation of a national policy. These costs are roughly estimated to add some 20 per cent to the total cost estimates presented here.

From Spain the cost estimate submitted by the Empresa Nacional de Residuos Radiactivos S.A. (ENRESA) for Vandellos 1, a 500 MWe gas cooled reactor (GCR), has been included in Table 1. The decommissioning of the facility is planned to be started in 1991. Stage-1 should be reached by 1995, after which the plant will be brought to Stage-2 by the year 2000. After a 25 year period of dormancy, final dismantling follows and is planned to take five years. Costs for radioactive waste disposal are not included in the estimate.

Based on studies for Vandellos 1 decommissioning and also on the previous NEA work, rough cost estimates have been made for other nuclear power plants in Spain. Based on the NEA report the assumption has been made that it costs about 21.9 billion pesetas [US$200 million] (in 1990 money value), on average, to decommission a 1000 MWe PWR and 31.9 billion pesetas [US$300 million] to decommission a similar size BWR. The results of the estimates are shown in Table 2.

The Swedish estimates (SW-1 and SW-2) in Table 1 correspond to Ringhals 1 and Ringhals 2 reactors, the former being a 780 MWe BWR of the Asea Atom second generation design, the latter a 920 MWe PWR of Westinghouse three-loop design, both at the same seaside location. The estimates have been prepared by the Swedish Nuclear Fuel and Waste Management Company (SKB) in 1986. Immediate dismantling (Stage-3) is assumed in the cost estimates.

The cost estimates for the reference plants mentioned have been used as a basis for estimation of the decommissioning cost of other Swedish reactors. The required adjustments have been made in proportion to material quantities. The results are shown in Table 3. Also included is the estimated cost of shutdown operations in the case where all Swedish nuclear power plants are shutdown in 2010. If the shutdown is phased over a five-year period, these costs will be only about 30 per cent of those indicated in Table 3.

From the United Kingdom estimates were submitted by British Nuclear Fuel plc (BNFL) and Nuclear Electric (NE). BNFL's estimate (UK-1) is for the total cost of decommissioning two Magnox stations, Calder Hall and Chapelcross, each having four 60 MWe reactor units. The stations are situated on or near the coast. The reactors are Britain's oldest commercial nuclear power plants and represent a fairly simple design with many features intended for ease of maintenance. Each reactor has four external boilers.

The planned decommissioning strategy assumes a three-stage programme. Stage-1 will be reached within 3 years after which the work will continue to bring the station to Stage-2 during the subsequent

10 years. The dormancy period after reaching Stage-2 is assumed to last for 60 years for Calder Hall and 90 years for Chapelcross. After the dormancy periods, the plants will be brought to Stage-3. About £215 M of the total costs are directly attributed to Calder Hall, about £275 M to Chapelcross. The balance, £345 M consists of allocated costs, including contingencies.

NE's estimate for a 2 x 219 MWe Magnox station (UK-2) is considered as representative of a typical Magnox station. Earlier Magnox reactors have steel pressure vessels and external boilers, whereas in the later designs the boilers are encased in the same concrete structure as the pressure vessel. The strategy for decommissioning is similar to that assumed by BNFL, although the time schedules are slightly different. Stage-1 and 2 are assumed to be reached in 5 and 10 years after shutdown respectively and a dormancy period is assumed to last for 90 years. However, the strategy is being reconsidered and a rescheduling of Stage-2 and 3 is being contemplated.

The strategy requires that the wastes from Stage-1 and 2 decommissioning are temporarily stored on site, pending the availability of a deep geological repository. At stations with steel pressure vessels, a temporary containment will also have to be built for the boilers, which will be disposed of at the time of Stage-3 decommissioning.

For comparison, NE has supplied estimates for advanced gas-cooled reactors (AGR) and a pressurized water reactor as well. The AGR estimate (UK-3) is considered representative for the average 2 x 660 MWe twin-unit facility with concrete pressure vessel which also enclose the boilers. The strategies assumed are similar to those explained for Magnox reactor decommissioning. The estimate for a PWR is applicable for the Sizewell B type station with 1155 MW electric capacity, assuming that Stage-1, Stage-2 and Stage-3 are carried out in a sequential and continuous way. However, the cost estimate for the decommissioning of a PWR is largely determined by conservative comparison with international cost estimates and is not a result of specific independent cost study.

The US estimate (US-1) has been prepared by the Pacific Northwest Laboratory for the Nuclear Regulatory Commission (NRC). The estimate corresponds to a four-loop Westinghouse PWR with a 1175 MWe power rating and is representative of large PWRs started up in the late 1960s to the early 1970s. It is assumed that the location is riverside and a cooling tower is installed. The original estimate was completed in 1978 and was significantly updated in 1986. In 1988 the corresponding data was used by the US regulators for developing the decommissioning fund rule. The US-1 cost estimate includes only those operations needed for dismantlement and decontamination for purposes of termination by the U.S. NRC of the reactor operator licence. The costs do not include structure removal and site restoration to a greenfield condition.

4.2 Comparison with other cost studies

1986 NEA report

In the 1986 NEA report on decommissioning, cost estimates were included from Canada, Finland, Germany, Sweden and the USA. The estimates were presented in US dollars of January 1984. Because of the marked fluctuations in exchange rates any direct comparison of the non-US estimates with the cost estimates submitted for this study would be meaningless. However, it is possible to re-convert the 1986 estimates to original currency units (national currency units as of January 1984) and compare these with the estimates presented in Table 1. Such a comparison is made in Table 4. The second column of the table shows the estimates presented in the 1986 NEA study.

The Canadian estimate in the 1986 study was submitted by Ontario Hydro and corresponded to the Pickering power station with 4 x 515 MWe of electric capacity. The cost estimate submitted to this study was prepared by the AECL and corresponds to a reference plant of 600 MWe CANDU reactor. The two estimates are based on separate studies and different facilities.

The Finnish estimates are for the same power stations and decommissioning strategies as the present estimates. However, after the publication of the 1986 report, revised decommissioning plans, including cost estimates, have been made for both stations. For the PWR station the revised detailed estimate is considerably higher than the earlier estimate, which was mainly based on cost studies from other countries and did not sufficiently take into account the plant-specific features. For the BWR station the difference between the estimates is small.

The German estimates in the 1986 NEA report were prepared in the 1970s by the Working Group "Decommissioning of Nuclear Power Plants" of the Association of German Utilities jointly with the NIS Engineering Company Ltd for a 1200 MWe PWR (Biblis A) and a 800 MWe BWR (Brunsbüttel). The estimates include for both reactor types the two options: immediate dismantling; safe enclosure of 30 years and delayed dismantling. The cost estimates for Biblis A were revised by the same estimator-group on the basis of recent knowledge and published in 1986. The German estimates in this report are the revised estimates.

The Swedish cost information in the 1986 report was based on a SKB study from 1979 and included estimates for the power stations mentioned in Table 1. The 1986 SKB study that has been used for the estimates of Table 1 was mainly an update of the earlier cost study and brought no significant changes to earlier estimates.

The case is similar for the present U.S. NRC estimate and the estimate presented in the 1986 report. Both estimates have been made by the Pacific Northwest Laboratory and are based on the same basic studies.

1989 NEA/IEA report

The third column of Table 4 shows the cost assumptions that were used for decommissioning in the 1989 NEA/IEA report on electricity generation costs [2]. In the report the estimates are presented in US dollars of January 1987, but for Table 4 they have been converted back to national currencies of the same base date.

In many cases the estimates of the 1989 report are similar to those presented in Table 1, while some exceptions also exist. However, the estimates of the 1989 report must be seen in their proper context. First, they represent potential future facilities with the start of operations sometime in the late 1990s and hence do not necessarily incorporate any plant- or site-specific data. Secondly, the report shows that the levelized costs of electricity generation are not sensitive to the assumption on decommissioning costs — particularly in relation to uncertainties in other cost components of electricity generation. Therefore, rough, sometimes perhaps purposely conservative, estimates may have been considered sufficient for that study.

Other recent estimates for decommissioning costs

A recent survey among US utilities brought up 68 estimates for nuclear power decommissioning costs [3]. Included in the survey are BWRs and PWRs of sizes ranging from 75 MWe to about 1300 MWe plus one high-temperature gas-cooled reactor (HTGR) of 342 MWe electric power capacity. All the cost estimates have been made during the last several years.

The cost estimates for large, over 900 MWe, facilities range from about US$115 million (1990) to about US$190 million (1990), the lowest estimate being for a 950 MWe PWR, the highest for a 1100 MWe BWR. For plants with smaller than 900 MWe power capacity the estimates are in most cases lower than US$150 million, the lowest cost, US$80 million (1989) being estimated for the HTGR mentioned above. In one case the decommissioning costs for a 845 MWe PWR are estimated as high as US$315 million (1989), but this estimate includes items which are not directly attributable to decommissioning. For instance, about 30 per cent of the total costs are caused by the on-site storage of spent fuel for the reason that no spent fuel repository is assumed to be available at the time of decommissioning.

The NRC estimate (US-1) in Table 1 falls at the lower end of the range of estimates for large reactors. The estimates included in the survey are based on varying assumptions and, as a closer look at the $315 million estimate shows, in some cases they may be rather different from those explained for the NRC estimate. Regulatory changes, industry experience and data refinements have influenced decommissioning estimates in several ways, possibly leading to different final impact in each individual case. Specifically, the following factors must be carefully considered and understood in any comparison: decommissioning to greenfield condition or alternative use, increases in costs for waste management and disposal, productivity shifts from changes in the size of staff and operations labour force, computer techniques that more precisely calculate costs, and the typically higher estimates resulting from conservative planning for government projects compared with commercial ventures.

4.3 Current cost experience with decommissioning

In the 1986 NEA report a review was presented of the current experience and developments in the field of decommissioning technology by that date. More than 20 projects were mentioned where nuclear power plants or fuel cycle facilities had been brought to various stages of decommissioning. Also included was a list of ongoing and planned decommissioning projects in OECD countries. Since 1986 some of these projects have been completed and some new projects have been started. Some of these are also included in the NEA Co-operation programme described in Annex 5.

Although the experience obtained in the ongoing and completed decommissioning projects is essential for the current planning of future projects, the costs of these projects are not directly comparable with the estimates produced in Table 1. First, the facilities are often very different from the present commercial facilities. Secondly, these projects often include a substantial component of research, development and testing of new techniques for decommissioning and, therefore, cannot be considered as examples of commercial decommissioning practice. However, the experience can be used to assess the efficiency of various techniques in realistic conditions and, hence, substantially reduce uncertainties in cost projections.

Cost information from four recent projects is included in Annex 2. Gentilly-1, a 250 MWe Canadian heavy-water moderated light-water cooled reactor was brought to "static state", a variant of Stage-1, during 1984-1986 and now is considered a waste storage facility. According to present plans, the dismantling will continue after a 52.5 year dormancy period aiming at release for unrestricted use of the facility.

The present work at the French Chinon-A2 reactor also aims at partial decommissioning. The present scenario consists of bringing this 250 MWe gas-cooled reactor to a "reinforced Stage-1" which will be followed by a dormancy period of at least 50 years. The reinforced Stage-1 is planned to be reached in 1992.

The German 100 MWe gas-cooled heavy-water moderated KKN Niederaichbach will be decommissioned directly to Stage-3. According to present plans all dismantling operations will be completed in early 1994 and the site will be released for unrestricted use.

Up to now the experience in these projects is limited to partial decommissioning only, so the costs realized in these projects cannot be directly contrasted with the estimates presented in Table 1. Since, however, all three reactors are representative of large power reactors, the cost experience from these projects goes a long way to substantiate the cost estimates that are made for individual decommissioning activities.

The fourth example discussed in Annex 2 is the Shippingport decommissioning project. The decommissioning work at this 72 MWe PWR plant started in 1985 and the site was released from regulatory control in 1989. The total cost of the decommissioning project was US$91 million, which was about 10 per cent less than estimated.

The relevance of the Shippingport experience for the decommissioning of commercial reactors in the future has been questioned on the basis that Shippingport still represents a fairly small power capacity and in this case use was made of techniques that may not be applicable elsewhere. For instance, it is said that one-piece removal of pressure vessel is likely to be impossible at most larger facilities and the utilities will probably have to pay much more for the waste disposal than the U.S. Department of Energy had to in this case.

In response to these arguments it should be noted, first, that the Shippingport plant had all the components of a large commercial power plant. In fact, a comparison of the physical size of the components and systems and the containment masses with other commercial reactor shows no significant differences. Table 5 gives some comparative data. The difficulty of the work and the associated time factors at power reactors will generally be fairly similar to those experienced at Shippingport.

Secondly, the feasibility of the reactor vessel one-piece removal compared to segmentation removal can be determined only from a cost benefit analysis that addresses site specific conditions. Physical size of the reactor pressure vessel is only one factor in such an analysis. Other factors include residual contamination levels and restrictions for transportation and available disposal options.

In broad term, the Shippingport project demonstrated that decommissioning can be planned, managed, and completed with the existing technology; and, that decommissioning costs can be estimated with reasonable accuracy.

5. FACTORS AFFECTING DECOMMISSIONING COST ESTIMATES

The purpose of this report is not to assess the validity of published decommissioning cost estimates. What follows is an attempt to provide answers to the question of why the estimates, such as those in Table 1, seem to vary so much. The discussion is focused on recognizing the most important institutional, technical and economic factors that can be expected to cause differences among cost estimates for different facilities, or even for similar facilities if they are situated in different countries or regions or otherwise subject to different boundary conditions as regards the planning of decommissioning projects.

The first thing to note is the general difficulty of making international cost comparisons that has been discussed at length in several earlier NEA reports (see, e.g. [2]). Since cost estimates have usually been made in national currencies they have to be converted to some common monetary unit before any international comparisons are possible. If, additionally, the estimates have been prepared in different years, further adjustments may be necessary.

Most often the comparisons have been made on the basis of US dollars and using normal currency exchange rates. However, the dramatic variations during the 1980s in the exchange rates between dollars and other currencies detract from the usefulness of such comparisons (see Figure 2). The exchange rate movements do reflect the variations in relative inflation rates between countries, but these alone cannot explain the recent huge fluctuations. For example, in January 1984 the rate of US dollar against the DM was 2.72, whereas in the beginning of 1987 it was already 1.94 and in early 1990 the rate was near 1.6. Since these fluctuations have no direct connection with the relative developments in unit factors such as labour rates or material costs in the two countries, dollar-based comparisons of German and US estimates made in successive years during this period are necessarily biased and likely to suggest tendencies which have no counterpart in reality.

The pertinence of exchange rates for international cost comparisons can also be questioned for methodological reasons. The exchange rates may give a fairly correct picture for the prices of products or services that are freely traded on international markets. In the decommissioning projects, however, a major part of the costs is attributable to domestic labour and, hence, depends on local labour rates. This part of the cost variations may be better analysed in terms of purchasing power parities of the currencies.

Different dates of estimates generate other conversion problems. Even if the inflation rates within the OECD area have recently been fairly low, the relative escalation rates may still have been more significant for some cost components. Country-specific indices for different cost components should be used for updating of the estimates, but the application of such indices would require very detailed information on cost structures.

Nevertheless, important as the conversion difficulties may be, they do not explain why separate estimates from the same country can also be different, even if made at about the same time. Indeed, there are often obvious reasons to expect that the estimates should be different. Numerous factors underlying the estimates cannot expected to be the same in each case.

These factors comprise the legal and institutional framework which controls all nuclear activities and, in particular, decommissioning, the national and local infrastructures and economic circumstances, the company strategies and the technical characteristics of the facility in question. Some of these are probably common to all nuclear plants in a certain country or region, while other factors are strictly specific to each facility or facility owner. Some of the factors the planner of a decommissioning project must take as given, while some of them may be under his control, at least to some extent. Since the factors are often interrelated it is very difficult, if not impossible, to discriminate between the effects of individual factors on an individual estimate. Therefore, in the following discussion of these factors the emphasis is on qualitative conclusions as the quantitative examples can not be taken to have precise general validity.

The logic of the presentation follows a top-down order. The factors that determine the external framework for the decommissioning project are discussed first. The discussion then goes step by step deeper into facility and project details which appear important for the understanding of decommissioning cost estimates.

5.1 Institutional framework

5.1.1 Nuclear energy policy and legislation

In all OECD countries the operation of a nuclear facility is subject to an operating licence, which sets the conditions that the facility must meet to be environmentally and socially acceptable. Normally the responsibility of the licencee does not end with shut-down of the facility but remains until the time the facility can be shown to present no risk to the population or environment, or when the plant area can be released to unrestricted use.

Even if the operating licence in some countries is given for a certain period of time there is no fixed technical lifetime for a nuclear facility. With varying amounts of refurbishment and upgrading it is possible, at least in principle, to maintain the plant in a condition that fulfills the requirements of the operating licence. In practice, however, the lifetimes depend on several economic and policy factors and may become quite different for different facilities.

Most of the commercial nuclear power plants within the OECD area are still young and are not expected to be shut down in the near future. However, a number of older, usually relatively small, facilities have been retired. In a few cases the operating licence has been cancelled or suspended because of technical deficiencies or other problems.

The plant owner's responsibility after termination of the operating licence has been defined in different ways in different countries. In some countries rules exist for how and when the decommissioning must be carried out. For instance, the rules may require prompt dismantling after shut-down and they may require that the site is returned to a greenfield state. In some other countries the practical approach and time schedule for the decommissioning can be decided by the plant owner and the limits of dismantling work depend on conditions set for free release of materials from regulatory controls.

In the former case the legal framework may significantly restrict the freedom left for optimization in the decommissioning planning and, therefore, indirectly affect the costs. In Japan prompt dismantling within 5 to 10 years after shut-down is considered a rule if not a necessity. In the USA the limit is strict but much more distant: dismantling must be completed within 60 years from the shutdown.

Without specific rules the plant owner may consider for himself how the responsibility for the safe containment of radioactivity can be best fulfilled. In this situation the plant owner is likely to choose the decommissioning strategy that best combines this objective with his company's other financial and strategic plans.

The policies for future electricity capacity expansion may affect the decommissioning plans and cost estimates in several ways. If the site continues to be used for energy production purposes, the final dismantling work may be postponed since the safe storage of the facility can be easily arranged as a part of other site operations. The personnel of the new facility may in this case take the responsibility for the surveillance and maintenance work needed to keep the plant in safe condition (Stage-1). Such an arrangement has been planned, e.g. in Italy.

On the other hand, in countries where land is scarce, efficient use of an existing nuclear site makes it desirable to dismantle the old facility before the construction of a new one is started. Prompt dismantling may be advantageous for the overall nuclear energy policy even if this might not minimize decommissioning costs in some cases.

In the case of a nuclear moratorium, the future availability of a skilled workforce for decommissioning project may be considered as a problem. In this case also rapid decommissioning may be considered reasonable even if costs would be higher.

Public attitudes continue to be critical of nuclear power use in many countries; one of the main concerns is with the destiny of the plants and wastes after the operations are terminated. Since public acceptance itself is sometimes a prerequisite, e.g. for the licencing of a waste disposal site, and since it may always become decisive through the political system, some of the decommissioning plans seem to have been built on more conservative assumptions than what the existing legislation would require. The unpredictability of the social and political context can lead to an increase in contingency allowances.

5.1.2 Regulatory framework

The radiological impact from decommissioning and subsequent waste disposal has been generally deemed to be small for both the workers and the general public. However, large amounts of radioactive materials need to be handled and a part of the work needs to be done under increased radiation levels. Regulatory rules are, therefore, needed to control the potential negative impacts and define what is acceptable to man and society. These rules — which are partly still under development in most OECD countries — may heavily influence the way the decommissioning work is done. Current developments in decommissioning regulations are reported in Annex 4.

Besides the potential radiological impact, decommissioning may cause other environmental and social effects in the plant vicinity. Regulations may exist to control these effects as well. For instance, it may be required, for visual reasons, that all the plant buildings are demolished, although this would not be necessitated by radiological reasons. Social considerations and public acceptance matters may particularly affect the strategy that is chosen for carrying out the work.

5.1.2.1 Exemption levels

Two basic issues that call for regulatory decisions are the criteria for release from regulatory control of materials and equipment and the long-term safety requirements for the final disposal of radioactive wastes. The establishment of detailed regulations requires that the judgement is made on what is socially and environmentally acceptable, remembering the long time span over which the impact needs to be regulated. Since the impacts from the practices may extend over national boundaries, efforts have been made to agree on common regulatory principles that would be applied worldwide. Basic principles to this effect have been proposed by the International Commission on Radiological Protection in 1985 (ICRP)[9].

Despite some consensus on the basic criteria, regulatory practices still differ widely in different countries. As to the exemption levels, i.e. the limits for radioactivity under which the materials or components can be exempted from regulatory control, further work has been done in various international organisations. Recommendations are included, e.g. in the IAEA Safety Series No. 89 [10] and the CEC Radiation Protection No. 43 [11].

In so far as the recommendations are given in terms of doses from the wastes they still leave plenty of room for different practical interpretations. Distinction may also be made between release for unrestricted use and partial release for specified use. Detailed site release criteria or specific exemption levels in terms of radioactivity contents have so far only been established in a few countries. For example, among the countries in Table 1, the UK alone has established an exemption level in regulations. Hence most decommissioning plans are still based on assumptions in this respect. As Table 6 shows, these assumptions may be quite different.

In general, the lower the exemption levels are, the larger will be the share of the waste considered as radioactive. If the decommissioning plan includes the dismantling of all components and buildings, the exemption limits mainly affect the waste handling and disposal costs, although they may also affect the way the dismantling can be done. If the dismantling is planned to be restricted to radioactive parts only, the impact on dismantling costs may be quite considerable.

The sensitivity of the decommissioning cost estimate to the assumed exemption level is different in different cases. First, the marginal costs of lowering the exemption level are generally the higher, the lower the level is, but the impact may be quite different in different countries, depending, e.g. on the level of waste disposal costs. Secondly, the extent and nature of the contamination is different at different plants, and, therefore, the same radioactivity limit may have different practical implications.

Low exemption levels give rise to measurement and verification problems. At Eurochemic Reprocessing Plant, the decrease of the exemption level by a factor of ten in the decommissioning of a final product storage led to a 50 per cent increase in the verification effort required for free release. That corresponds to a 4.3 per cent increase in the total costs. There the surface contamination limit for beta-gamma radiation was changed from 4 Bq/cm^2 to 0.4 Bq/cm^2 and the limit for alpha radiation from 0.4 Bq/cm^2 to 0.04 Bq/cm^2. If the limits are set very low, it may become less costly to consider some wastes as radioactive and treat them accordingly than to try to prove that they meet the requirements for free release.

In some estimates the regulatory indeterminacy is accounted for in contingencies. In the Finnish estimate for a BWR plant decommissioning an allowance of about 7 per cent has been made because of the uncertainty in the applicable exemption levels.

Because of the regulatory uncertainties and verification problems recycling or re-use considerations have mostly been ignored in the present decommissioning cost estimates. Quite a lot of the material obtained from dismantling a nuclear facility could, in principle, be re-used, but the cost benefits should be weighed with the costs of licensing and carring out such operations. In the estimates of Table 1 no allowance has been made for the scrap or residual values of the components or materials.

5.1.2.2 Criteria for radioactive waste disposal

In current technical language a distinction is usually made between low-level, intermediate-level and high-level radioactive wastes. Qualitatively, the distinction between low and intermediate levels is sometimes described on the basis that no shielding should be required in handling and transportation of low-level wastes; heat generation might be the distinctive feature for high-level wastes [12]. An additional distinction may be made on the basis of the need for long-term isolation from the biosphere. Actual quantitative definitions for waste categories vary widely from one country to another, and at present there are no generally agreed definitions for the terminology. The present situation is summarized, e.g. in a report by CEC [13]. Two examples for waste classification are also shown in Tables 1.3 and 1.7 of Annex 1.

Quite understandably, there are no standard solutions for disposing of wastes of different classes, either. Therefore, the classification as such has little meaning in international comparisons of waste management and disposal costs.

Nevertheless, differences in regulations may lead to important differences in disposal costs. Although the ICRP recommendations are largely accepted as the basic radiation protection criteria, there seem to be significant differences in the proposed technical criteria derived from them and in their planned site-specific application (still under preparation in many countries). Accordingly, different technical disposal solutions are envisaged in different countries, the alternatives extending from sea dumping or shallow-land disposal to deep underground repositories [13].

Where waste acceptance criteria for certain repositories already exist, these may be decisive for the disposal costs. In the UK, for instance, BNFL plans to use the Drigg facility for low-level wastes (not exceeding 4 GBq/t alpha activity or 12 GBq/t beta-gamma activity) and estimates that the average handling and disposal costs for them would be about £650/m³, whereas for intermediate-level wastes to be disposed of at the proposed Nirex facility the estimated costs would be £2 700/m³ on the average. The change in the classification limit so as to move 15 per cent of the low-level wastes into the intermediate-level category would increase the total decommissioning costs by some 5 per cent.

A similar cost ratio for shallow land burial and geologic disposal is assumed in the cost estimate for decommissioning the Eurochemic reprocessing plant at Mol. There the costs for intermediate storage and final disposal of low-level wastes were estimated at BF 115 000/m³ whereas for intermediate-level wastes the estimate was BF 675 000/m³.

Indeed, a major factor affecting waste disposal costs is whether the option of near surface or shallow-land disposal is assumed to be available for certain low-level wastes or not. If that option exists, the acceptance criteria for such repository become important for the costs, whereas in cases where all radioactive wastes are bound for geologic disposal in any case, the classification of wastes may not have such importance.

This is illustrated by the situation in UK. Whereas BNFL plans to use the option for shallow land disposal of low-level wastes at Drigg (estimate UK-1), the NE assumes geologic disposal of all radioactive wastes at the Nirex facility (estimate UK-2). For NE the cost for disposing of the intermediate-level and of the low-level wastes differs by a factor of 1.4 on the average, while for BNFL the difference is a factor of 4.

5.1.2.3 Occupational dose limits

The regulatory dose limits for the workers at nuclear facilities are similar in all OECD countries and correspond to the ICRP recommendation of 50 mSv/year. However, the ALARA principle requires that the exposures be kept as low as reasonably achievable. The practical target levels may, therefore, be much lower than the regulatory limits. In Table 1 the estimate from BNFL (UK-1) is based on the dose limit assumption of 20 mSv/year. Nuclear Electric has used one tenth of the limit as a target for workers' average dose (estimates UK-2, UK-3, UK-4). The other estimates of Table 1 are based on the application of the current regulatory limits.

The international recommendations concerning the occupational limits are being reviewed and a decrease of the limit is likely. The new limit may have implications for the plans that assume dismantling soon after shutdown, whereas in the cases of delayed dismantling after several decades the dose limits are likely to be less restrictive.

5.1.2.4 Other regulatory considerations

Transport regulations affect the transportation costs. Since, however, waste transportation is a relatively small item in the total decommissioning costs, regulatory differences are not likely to cause major differences in total decommissioning cost estimates. If the wastes can be disposed of on-site and public routes do not have to be used at all, considerable savings may, however, be achieved.

The costs during the dormancy period before final dismantling are usually estimated assuming that a minimal staff is needed to provide the safe storage. Regulatory rules might affect these costs. However, if the costs of safe storage turn out very high, reconsideration of the whole decommissioning strategy becomes likely.

In general, estimates differ in their ways of handling the future expected developments in regulations. For example, in Finland the cost estimates that are used for determining the funding requirements for decommissioning shall, according to the legislation, be prepared using current regulatory requirements as the basis. In the USA, on the other hand, some utilities — having learned from the recent history of changes in plant regulations — provide contingencies for the future regulatory changes.

5.1.3   Financial responsibility

In many countries rules already exist for the financing of the future decommissioning work (see Annex 3). In some countries a fund has to be established to collect money from the operating revenues during the lifetime of the facility. This fund — or at least the amounts of money that are set aside — is controlled by authorities. The authorities may require that detailed plans and cost estimates are prepared for decommissioning before the operating licence is granted. Further, the stipulations may require that the estimates are made on a conservative basis.

The money set aside may stay within the firm, but the authorities still have a control on the system. In the USA, for example, the public utility commissions, which regulate the electricity prices, may decide how much the utilities are allowed to collect from their customers for decommissioning. Some utilities have complained that the present charges will be insufficient to cover the costs of the future decommissioning work [14]. In particular, criticism has been directed against using the NRC estimate (estimate US-1 in Table 1) as a basis for rate-making decisions, since many utilities believe this is far too low. Utilities' own estimates are generally higher, but they also show a considerable variation [3].

In some countries both the responsibility and the mode of financing is left to the facility owner's own decision. Even though general rules for decommissioning may have been set out in the law and regulations, it is considered that there is no value in making detailed plans for decommissioning many decades ahead.

Financing methods are not likely to have a significant direct impact on decommissioning costs. Nevertheless, they may affect the decommissioning strategy and, hence, indirectly the costs. If a high rate of interest can be earned with the funds collected for decommissioning, it may be advantageous to postpone the final stages of decommissioning. In this respect the situation may vary from country to country.

Other implications of financing rules on decommissioning plans will be discussed later in Section 5.3.

5.2 Infrastructure development

5.2.1 Technology

The experience from current decommissioning projects and repair and maintenance works at operating nuclear facilities forms the basis for planning future decommissioning work and for estimating the costs thereof. Technical approaches to decommissioning are basically similar throughout the world. However, the growing number of decommissioning projects is creating expanding market opportunities for new techniques and equipment and, hence, incentives for increased research and development efforts. Therefore, it is likely that technical plans will be rapidly evolving as new technical means become available. At the same time, because of the growing commercial interests involved, detailed descriptions of the techniques and the precise cost data may often become confidential in nature, thus making reliable comparisons difficult.

The differences between countries as regards access to new technologies or the domestic availability of certain equipment give rise to differences in cost assumptions. In some cases different regulations (or different assumptions concerning the future regulations) may cause differences in technical approach and, consequently, in costs. Different assumptions concerning acceptable occupational doses might give rise to differences in the choice between the use of direct hands-on methods and remote-controlled techniques. Certain techniques for the disposal of low-level radioactive waste, e.g. the shallow land disposal, may be prohibited in some countries although they are being used in some other countries.

Generally, the estimates may differ in their relation to the expected future development. Future technology is likely to rationalize, optimize and simplify decommissioning and, therefore, lead to cost savings. On the other hand, the new technology and knowledge may sometimes reveal previously unknown technical problems which may cause additional costs. In most cases, however, the cost assumptions correspond to present circumstances and to the present status and availability of technologies.

5.2.2 Availability of disposal facilities

Decommissioning to Stage-2 or 3 gives rise to a considerable amount of radioactive waste. At nuclear power plants most of this is usually low-level waste which can be handled without shielding and for which the need for isolation is of short-term character. Some of the activated reactor materials will need more substantial packaging and isolation. The waste from certain fuel cycle facilities may additionally contain transuranic waste which needs very long-term isolation from the biosphere.

Facilities for the disposal of low-level radioactive waste exist or are under construction in several countries but, in most cases, new facilities will be needed to accommodate the decommissioning waste. The current situation is summarized in Table 7. In some countries plans for disposal facilities exist while in other countries decisions are still needed at the policy level before actual design of repositories can be started.

The status of development in this area affects the cost estimates both directly and indirectly. It is normally cheaper to expand the capacity of an existing facility than to build a completely new one. Another direct effect may be seen in contingencies which are likely to be high in the cases where large uncertainties still exist on the applicable disposal principles or on the schedule on which facilities are expected to become available. Indirectly the availability of disposal facilities affects the costs by affecting the choice of the strategy, since it is reasonable to postpone the dismantling until facilities for disposal are available.

The perception of the future waste handling costs may also be different depending on whether the plant owner is responsible himself for the construction and operating the disposal or whether the wastes will be taken care of by a government facility with a fixed price for the service. A large centralized facility may benefit from the economies of scale but it may lead to greater transportation distances than a facility that serves one or a few customers only. If the plant owner is responsible for the construction and operation of the disposal facility, his optimum decommissioning strategy may also turn out different from the case where the plant owner is subject to fixed costs and to fixed conditions of acceptance.

5.2.3   Human resources

Under a nuclear moratorium, concerns about the availability of a skilled workforce may strongly affect the choice of decommissioning strategy, as was explained earlier. Resource factors may be important for costs in all cases where, for one reason or another, a shortage of skilled manpower is anticipated. As soon as the technology-specific expertise is lacking, extensive training programmes will be needed which add to the labour costs.

5.3   Decommissioning plans

5.3.1   Status of the decommissioning plans

A considerable amount of experience exists for decommissioning small research, test and demonstration reactors and for a few early prototype power reactors. The number of power reactor shutdowns is likely to remain fairly small at least for a decade or so.

It is understandable, therefore, that decommissioning plans for large commercial nuclear facilities may still be very preliminary. In some countries the preparation of detailed decommissioning cost estimates for the operating commercial reactors is, in fact, considered premature, as no practical decommissioning work is expected to be done until sometime in the 2010s. In these countries, generic cost studies like those made by the UNIPEDE and NEA have been considered sufficient for the present planning of financing and R&D policy.

However, as was explained above, in some other countries the utilities are obliged by law to prepare detailed decommissioning plans, including cost estimates, before the startup of the nuclear facility or at least during the early years of operation. These plans are usually plant-specific and may go to the level of detail which would enable the start-up of the practical decommissioning project whenever this might become necessary.

Generic estimates may, of course, be based on detailed data and assumptions and they may include sensitivity cases to allow for plant-specific variations. Even so, they may not be able to identify and illustrate with all the interrelationships of technical and other factors that affect the cost estimation for every specific case.

Most of the estimates in Table 1 have been prepared for specific power plants. The US estimate is considered representative of a typical PWR plant in the USA. It is based on a common design but replaces the site-specific variations by generic assumptions. A similar approach has been used in the Canadian, German and Japanese estimates.

The technical level of the decommissioning plans may have consequences in the cost estimation. With detailed characterization of the decommissioning project it is possible to remove major uncertainties concerning the basic data such as activity inventories, radiation levels and waste amounts. Although knowledge of such details may not resolve all uncertainties, it may justify a smaller contingency allowance than would be appropriate in the case of less detailed plans. Furthermore, the optimization of a decommissioning plan is possible only if reasonably detailed plant-specific information already exists.

The level of detail also affects cost comparisons. Details indicate what is covered by the estimate and what is the status of technical plans and, therefore, facilitate comparison and assessment of cost estimates.

Comparisons between plans which are very different in their degree of detail may not be useful or practicable at all.

5.3.2 Scope of the estimate

Decommissioning has been defined to mean all those activities that begin after operations have ceased and are intended to place the facility in a condition that provides protection for the health and safety of the decommissioning worker, the public and the environment [6, 1]. Broadly speaking, most estimates for decommissioning accord with this definition. A closer look, however, reveals differences at two levels. First, there are differences in the planned scope of the work to be done; secondly, even if the scope of work is similar, the inclusion of certain cost items in the decommissioning cost estimate may be arbitrary (see Table 8).

In Table 1 defuelling costs are included only in the Canadian, the Swedish and the UK estimates. In the case of Magnox reactors and advanced gas-cooled reactors (AGR) defuelling is considered a major effort in the post-operational period with a 5 to 10 per cent share of the total costs. In the Canadian estimate, costs for shutdown operation including fuel removal is estimated 2 per cent of the total costs. Reprocessing of the final core loading, however, is not included in any of the estimates shown.

Variations appear likely as to the extent the long-term research, development and planning costs are included in the individual estimates. In the estimates of Table 1 account is taken of the costs that can be directly assigned to the project such as those for preparation of plant inventories and work packages for decommissioning. The share of these costs is reported to be between 1 and 2 per cent of the total.

The extent to which the facility owner's management and administration costs are included is often arbitrary. The case where the decommissioning project is just a small part of the company's total activity may be quite different in this respect compared with the case where the company's existence is tied in with the facility for which the decommissioning cost estimate has been made.

The significance of owner's personnel costs, in general, is illustrated by a Swedish case study [15]. In the study it was estimated that the shutdown operating costs — during the period between the plant shutdown and the beginning of actual decommissioning activities — could account for some 10 to 15 per cent of the total cost if all the nuclear power plants were shutdown at the same time. The costs were found to be strongly dependent on the length of that period and on the number of technical and administrative staff needed. The staff requirements, on the other hand, depend on the extent of other activities at the same site. If the decommissioning activities were optimally scheduled, the shutdown operating costs would be only one third of the costs in the special case considered. Additional staff costs may arise in every case where decommissioning activities have to be delayed because of external constraints.

There are differences also in how the management and administration costs are calculated in the estimates. In some estimates the required number and kind of such personnel is explicitly estimated while in other estimates they are included as indirect costs through factors applying to the direct costs for technical labour. These differences make direct comparisons of the numbers of personnel difficult.

Perhaps the most important difference between the scopes of the estimates is in the understanding of what constitutes decommissioning. As shown in Table 8, the estimates from Canada, Germany, Spain, Sweden and UK include all the costs that are necessary to bring the site to a greenfield condition. The Japanese estimates include the dismantling of the buildings except some auxiliary facilities and the base-mats of the reactor buildings. In the Finnish estimates and in the US estimate the costs are considered only for dismantling the parts considered to be radioactive and for the corresponding waste management and disposal.

Differences in the scope of work may be of regulatory origin. Sometimes the present requirements may not be precise enough in this respect but the greenfield state is used as a conservative assumption. In the USA it is estimated that the costs may be up to 40 per cent larger in the case of complete dismantling of a nuclear power plant than in the case of dismantling limited to radioactive parts only. In the Swedish estimate 15 to 20 per cent of the total dismantling costs correspond to the costs for demolition of buildings considered as non-radioactive and the costs for site restoration.

Even the definition of the "greenfield state" is somewhat arbitrary. Usually only a part of the underground concrete structures are planned to be demolished but the assumption on the exact depth varies.

In the cases where decommissioning is limited to radioactive parts only the total costs are sensitive to the regulations concerning the exemption limits, as was discussed earlier. Differences in these limits together with differences in the extent the plant structures are contaminated cause variation in the actual scope of decommissioning work.

The Japanese estimates in Table 1 do not include the waste disposal costs as the regulations in this respect are not ready yet. According to preliminary Japanese reviews, the disposal costs could add some 20 per cent to the present cost estimate, but they are likely to depend on the applicable exemption level. The Spanish estimate in Table 1 is in a similar situation. On the other hand, the Canadian estimate includes the costs for removal and disposal of replacement pressure tubes, and disposal of radioactive wastes from 30 years of plant operation.

Some differences are evident in the allocation of waste disposal costs. They may arise because the distinction between plant operating wastes and decommissioning wastes may be vague, e.g. for the activated core components. Furthermore, if the decommissioning wastes are to be disposed of in the same facility as the plant operating wastes, the allocation of fixed construction costs to the two types of wastes is somewhat arbitrary.

5.3.3 Decommissioning strategy

As explained in Chapter 2, there are two principal approaches to the decommissioning of a nuclear facility. Either all the dismantling work is done shortly after the shutdown or the dismantling proceeds in stages between which there are periods of varying lengths during which the radioactivity in the remaining parts of the facility is allowed to decay to lower levels. In practice the choice may be effectively constrained if the regulations call for rapid dismantling or the lack of facilities for disposal — or, sometimes, the lack of applicable rules and regulations for disposal — make it desirable to postpone the dismantling work.

In the absence of binding external constraints the choice of the strategy will be based on a multitude of technical and economic considerations. The need to minimize waste quantities and radiation doses to workers may speak for delayed dismantling. In Figure 3 the normalized dose rate in a shutdown boiling water reactor (BWR) is depicted as a function of time. It is seen that during the first several decades, every 18 years a decrease about by a factor of ten is obtained in the total activity and the implications of the decay in the radiation levels at some parts of the plant may be even more important. Reduced radiation levels facilitate practical work; reduced activity inventory may imply lower amounts of radioactive waste.

In addition, delayed dismantling may help in financing and offer more time for planning and developing the decommissioning programme. It may also be reasonable because of the company's overall nuclear waste management strategy.

All these factors may produce savings to the plant owner. However, these direct or indirect savings must be balanced against the additional control and surveillance costs arising during the storage periods.

Another important factor is the availability of resources and infrastructures: during a prolonged storage the workforce and local services that had been established during the plant active operating period may be lost.

According to the 1986 NEA report the direct costs were lower for immediate dismantling than for the delayed strategy in the cases studied for USA and Germany while for the Canadian case the opposite was found true [1]. The differences in undiscounted costs were not very large; with discounting the delay becomes, of course, favourable even at fairly low discount rates.

The assessment of the alternative strategies is likely to depend on the facility type and characteristics and on local and national circumstances. In Finland one of the nuclear utilities opts for immediate dismantling, the other for delay. In the latter case one basis for the decision is the fact that the maintenance and surveillance during the storage period can be easily arranged because of other waste management activities at the site.

For gas-cooled reactors long storage periods may be especially advantageous. Because of the complicated internal structure of the pressure vessel a significant advantage may be obtained by having workers enter it during the dismantling. A storage period of about one hundred years will sufficiently lower the radiation level inside the reactor to make the hands-on work feasible (Figure 4). On the other hand, the dismantling of a light-water reactor pressure vessel by remote-control equipment is technically much simpler. Therefore, a hundred-year storage before dismantling produces considerably more benefits in the case of gas-cooled reactors than it would do in the case of light-water reactors. For the latter reactor type, the assumed delay is usually only several decades, and the pressure vessel will be dismantled by remote-controlled equipment in any case.

Strictly economic considerations may not be decisive. According to one estimate, delayed dismantling could decrease the collective dose to the decommissioning workers by some 20 to 30 per cent [3]. Related to the ALARA principle this would be a strong argument for delayed work. Concerns about availability of a skilled work force and the prevailing public opinion may push the decision in another direction. Depending on the funding method and the firm's current situation the financial argument for delaying may have stronger or weaker emphases in different countries or for different utilities. Therefore, published decommissioning cost estimates seldom represent the results of simple minimizations of undiscounted costs.

5.4 Facility characteristics

In the following parts the discussion is restricted to nuclear power plants only. Some considerations specific to decommissioning of nuclear fuel cycle facilities are discussed in Annex 1.

5.4.1 Reactor type

The cost estimates obtained for this study (Table 1) correspond to five different reactor types. The difference between the two light-water reactor types, the PWR and the BWR is relatively small, set alongside the difference from gas-cooled reactors or heavy-water reactors. Typical plant designs corresponding to the reactor types discussed in this report are illustrated in Figures 5, 6, 7, 8, 9.

The main structural differences between PWRs and BWRs are in the containment design and certain pieces of equipment. In PWRs the water flowing into the reactor is kept separate from the water (steam) entering the turbine, which necessitates the use of heat exchangers (steam generators) between the two water circuits. In BWRs the steam from the reactor is directly fed to the turbines. In PWRs the whole primary circuit — including the steam generators and pressurizers — is located within the containment, whereas the BWR containment can be made more compact. The BWR pressure vessel, on the other hand, is usually larger in volume than a PWR vessel for the same power capacity.

The structural difference means that in PWRs the contaminated cooling waters are normally confined within the containment while in BWRs some contamination may be carried by the steam up to the turbines. This difference may lead to greater volumes of radioactive waste from BWR plants [1]. However, the total activity inventories at PWR and BWR plants are often fairly similar and the differences in plant type may be overridden in importance by the specific design characteristics.

The comparison of cost estimates given in the 1986 NEA report for PWRs and BWRs in each country tend to indicate a higher cost for BWR decommissioning than for PWR decommissioning. The examples of Table 1 are less conclusive in this respect. In Sweden the BWR dismantling is estimated to cost about 20 per cent more than the dismantling of a similar size PWR. The difference is mainly due to dismantling of systems. Another study made in the USA suggests that in similar circumstances and for similar size the decommissioning costs would be about 25 per cent higher for BWRs [16]. On the other hand, the difference between the Japanese estimates for PWRs and BWRs of similar sizes is very small and in the case of Finland the result of the comparison is reversed. However, in the Finnish case the PWR concerned is a modified VVER-440 design having considerable differences from the Western PWR designs used in other OECD countries; also the decommissioning strategies in the Finnish cases are different.

In principle the main structural difference between light-water reactors and gas-cooled reactors is the gas-cooled reactor's large graphite moderator; about 4 000 tonnes of graphite is needed in one Magnox core. In addition, because of the natural uranium fuel, Magnox reactor cores are considerably larger in size than a PWR or BWR core corresponding to a similar power capacity. In practical designs the differences between the Magnox reactors, the most of which have been built before 1970, and the present-day light-water reactors are obvious in many other respects as well. For instance, the Berkeley Magnox reactor has eight primary coolant loops and steam generators, while three to four loops are normal in a Western PWR. Some of the Magnox reactors have a concrete structure which encases both the pressure vessel and the steam generators and which also acts as a biological shield.

Altogether, the waste amounts arising from decommissioning a Magnox reactor plant are much larger than those from a PWR or a BWR. It has been estimated that the amount of radioactive waste from a typical twin-unit Magnox plant is more than threefold in relation to the wastes estimated to arise from decommissioning the Sizewell B reactor (see Table 9). The difference is even more important when one notices the differences in net power capacity. The size aspects will be discussed in the next chapter.

Many points on Magnox reactors are common to advanced gas-cooled reactors (AGR), though quantitative differences between an AGR and a PWR or BWR are not so striking as between a Magnox reactor and a PWR or a BWR. In AGRs the graphite moderator is smaller in volume than that of a Magnox of similar capacity. AGRs have pre-stressed concrete pressure vessels. The waste amounts arising from decommissioning an AGR are larger than those from a PWR or BWR corresponding to a similar capacity but are smaller than those from a Magnox reactor plant of similar capacity.

Present CANDU heavy water reactors also have larger cores than LWRs of similar sizes. Another difference from LWRs is the pressure tube concept: the CANDU reactor assembly consists of a cylindrical low pressure stainless steel calandria vessel, penetrated by a number (380 in the case of a 600 MWe CANDU) of fuel channel assemblies. Fuel channel assemblies run through the calandria horizontally and contain the bundles of natural uranium. The calandria is filled with heavy water moderator. The heat of fission generated within the fuel is removed by pressurized heavy water coolant. The removal, storage or detritiation of the heavy water is a major cost item at CANDU decommissioning. The costs for dismantling of the complex reactor core and primary coolant systems may occupy a relatively larger share in the cost estimate for CANDU decommissioning than in the cases of LWR decommissioning.

5.4.2 Reactor and plant size

The electric capacities of the power generation facilities considered in Table 1 vary from 480 MW (UK-1) to 1470 MW (FI-2) (total design gross capacity of plants covered in the estimate). The UK-1 estimate is based on 2 small 240 MW Magnox stations with four 60 MW reactor units each. The FI-1 estimate is based on a power station with two 735 MW BWRs. The largest single unit is a 1204 MW PWR (GER-1,2). It is evident that both the size and the number of units at the same plant affect the costs, but the relation between these factors is not simple. The dismantling of a single unit plant is likely to produce less waste and require less manpower effort than the dismantling of a multi-unit plant of the same total capacity. On the other hand, the dismantling of two similar units certainly does not cost twice as much as the dismantling of one unit alone because of collateral costs and sometimes also because the units may not be completely independent of each other as some auxiliary facilities may be common. It would, therefore, be artificial to try to deduce the dismantling cost per unit from an estimate prepared for the whole plant.

In the 1986 NEA study the individual cost estimates were linearly scaled to correspond to the same facility size of 1300 MW. In most cases the estimates were originally for a smaller plant size. Because every decommissioning project entails fixed costs that are independent of the plant size, the scaling was, in these cases, thought to lead to conservative estimates.

It is fairly obvious that for the same reactor type and the same basic design and vintage the decommissioning costs are increasing for increasing power capacities. However, the assumption of a linear relationship between capacity and cost is likely to exaggerate the effect. For small capacity differences there may be no significant effect on costs at all. For larger capacity differences the effect may be hard to single out from, e.g. the vintage effect, since the larger capacities usually correspond to newer vintages as well.

Another way to look at the effect of the physical size of the facility on the costs of decommissioning is to compare the costs with the waste quantities produced from decommissioning. Table 10 shows the total masses of radioactive waste that are estimated to be generated in the decommissioning of the plants mentioned in Table 1. In the last column on the right the waste quantities are proportioned to the power capacities.

It is readily seen that per megawatt of power capacity the amount of radioactive waste from a Magnox reactor may be one order of magnitude higher than the corresponding amount for light-water reactors. Even for light water reactors the specific waste amounts differ considerably; another fact suggesting that no simple relationship exists between the power capacity and the physical size of radioactive structures that will have to be dismantled during decommissioning.

In Table 11 the estimated decommissioning costs are proportioned first to the power capacities, then to the quantities of decommissioning waste. In general, the specific costs per waste quantity show less variation than the costs per capacity. The comparison is particularly interesting for the Magnox reactors: while the decommissioning costs per power capacity are one order of magnitude higher for Magnox reactors than for light-water reactors, the costs per waste quantity are fairly similar for all reactor types.

It should be noted that for final disposal costs the volume of treated radioactive waste to be disposed of is an influencial factor. For evaluation of waste production costs, bulk volumes of radioactive metal, concrete and/or other materials is the influencing factor.

To summarize, the decommissioning costs are likely to increase with the increasing physical size of the facility. However, the amount of waste arisings is a better indicator of the size than the power capacity alone.

5.4.3   Design and vintage

Even plants of the same type may be significantly different in their design and site layout. Among PWRs, the Soviet design VVER-440 pressurized water reactor has six main loops with horizontal steam generators while a Westinghouse pressurized water reactor has three or four loops with vertically placed steam generators. Although the heat transfer areas of the steam generators might be similar per power capacity in both types, the total steel amounts are different. Moreover, the number and position of the steam generators affect the containment size and layout.

For BWRs, in particular, several containment designs exist as can be seen in Figure 10. The main circulation pumps are sometimes located inside the pressure vessel, sometimes outside. The design differences may affect both the way the dismantling of the containment and the pressure vessel can be done and the waste quantities. Design differences in PWR containments are shown in Figure 11.

In comparison with the early vintages, the development in structural materials and in computational capabilities of structural analysis has enabled reductions in the basic material requirements at power plant construction. Another development with newer vintages is related to the growing emphasis on maintainability and cleanliness of plants. Good accessibility to plant equipment and the pursuit of low radiation exposures also facilitate the decommissioning work. The maintainability requirements have also influenced the choice of materials so as to reduce the accumulation of activation products that severely increase the radiation levels. This may have an effect on the work efficiency and on the final activity inventories and amounts of radioactive waste. On the other hand, the increase in requirements for safety-related equipment may have increased the amount of structural components. In the Swedish study it was estimated that newer BWR units would require an additional four months for dismantling due to increased amounts of redundant and safety related equipment. That would lead to additional dismantling work and waste amounts to be disposed of.

According to French studies new vintages will probably be cheaper to decommission than the earlier designs [17]. Another comparison from US gives evidence that for PWRs the vintage effect may even offset the size effect, i.e. the newer designs are cheaper to decommission in spite of their larger power capacity [3]. For BWR decomissioning in the USA, however, the costs seem to correlate positively with power capacity.

5.4.4   Plant location

The location of the power plant may have some implications on the decommissioning costs. The siting of a nuclear facility in an area with high seismic activity — as is generally the case in Japan — means increased requirements of construction materials. This leads to more extensive demolition work and increased amounts of waste, even if a part of the base materials might be left in the ground.

The plant location is associated with the cooling system used. For instance, the cooling towers that are customary in Central European river-side locations are not needed at northern sea-side sites in Sweden and Finland.

As there will generally be a few waste sites or possibly only one per country, the waste transportation distance depends on the plant location. However, the impact of the distance on total waste management and disposal costs is likely to be relatively small since the transportation costs are usually dominated by the fixed costs of equipment.

Apart from the fact that labour costs and costs of equipment are different in different countries they are also dependent on the local availability of work force and services. In remote areas a premium may have to be included in wages to attract the work force. If the whole decommissioning is to take place soon after the power plant shutdown, the infrastructure that has been necessary for the power plant operation may be sufficient to support most of the decommissioning work as well. If dismantling is postponed, the establishment of necessary supporting services may cause additional costs.

5.4.5 Operational history

The extent of plant contamination naturally affects the decontamination and dismantling work as well as the quantity of radioactive waste arising. Depending on the operational history even similar plants may be quite different in this respect.

The length of the operating lifetime normally affects the inventory of activated waste and is likely to correlate positively with the extent of plant contamination also. However, for the water-cooled reactors the amount of contaminated waste is more sensitive to the number and kind of various leaks and spills at the plant. First, the contamination of the cooling water systems largely depends on the integrity of the fuel rods. Second, the integrity of the cooling water systems determines the cleanness of the plant. A major accident, of course, may considerably increase the quantity of waste arisings and the amount of effort needed in decomissioning.

Decommissioning cost estimates can truly reflect only the part of the operating history that has so far materialized. For the remaining part of the operating life assessments can be based on extrapolation, but the possibility of serious incidents or accidents can only be judged or statistically inferred. In practical cost estimation it may be accounted for in the judgement of contingency allowance.

5.5 Technical approach

The main activities in decommissioning work were described in Chapter 2. The basic approaches today are mostly similar and based on current available technology but some details and choices in the technical plans may have non-trivial cost implications.

One thing to be decided is the extent of decontamination activities. In strategies where the main parts are dismantled after several decades of storage under surveillance, the natural decay of radioactivity during the storage period may reduce the need for decontamination to simple flushing and removal of loose surface contamination. For dismantling shortly after plant shutdown the need for decontamination is related to the selection of working methods at dismantling. With extensive decontamination the need for remotely controlled equipment is diminished; however, the optimized level of decontamination will depend on ALARA considerations. The facility differences lead to different optima.

Another major choice is whether the pressure vessel is removed intact or whether it is first cut in pieces. At Shippingport the pressure vessel was shipped intact from the plant to the disposal facility. This is estimated to have led to savings of one year in project time and about $6.5 million in costs (total decommissioning costs were $91.3 million). Similar approaches may be followed at other plants where site specific conditions and cost-benefit analysis indicate that a one-piece removal is cost effective. At the Finnish Loviisa station one-piece removal is estimated to lead to savings of about 4 per cent. However, most decommissioning plans for large commercial reactors are currently based on segmenting the pressure vessel into pieces.

Differences are obvious in the planned waste disposal systems. In some countries at least the low-level waste from reactor decommissioning can be disposed of at shallow land disposal facilities whereas in some countries geologic disposal is planned for all radioactive wastes. The disposal manner affects the costs: the effect for low-level waste disposal is illustrated by Figure 12 which shows a comparison that was recently made by the CEC on this subject [18]. It should be noted that the facilities referred to in this figure may not be intended for disposal of decommissioning waste and the cost estimates indicated here may, therefore, be different from those that are used in estimates of Table 1. For instance, the costs for disposal of the decommissioning wastes from Swedish reactors correspond to building an extension to the SFR1 facility and are, therefore, estimated to be lower than those in Figure 12.

The possibility for in-situ disposal of the main plant components could substantially lower the costs [19]. In UK the method would consist of mounding over the reactors with sand; in Canada the idea is to bury the reactor into a pit underlying the site.

Plant-specific characteristics give rise to further variations in technical plans. For instance, in the first Magnox reactor stations the boilers are situated in separate chambers outside the pressure vessel containment. After Stage-2 a temporary storage will have to be established for the boilers before their shipment to disposal site at the time of Stage-3 decommissioning operations.

5.6   Economic factors

5.6.1   Labour costs

In spite of the planned use of robots and other automated equipment decommissioning is usually a labour-intensive project. Variations in wage assumptions and labour productivity may, therefore, have large effects on any comparisons. For instance, the recent statistics of the International Labour Organisation (ILO) for the countries included in Table 1 indicate a variation by a factor of about 2 in average hourly wages of workers in manufacturing and constructing industries at the end of 1980s[1] [20] (see Annex 6). The ILO statistics are for the gross wages to workers (before tax deduction) and, hence, do not represent the total labour costs, which also include various contributions to health care and insurance schemes and other possible benefits paid by the employer. Besides, they need not be consistent with the estimates of Table 1 which are based on assumptions of future costs and may include allowances for future cost escalation.

Labour cost assumptions used for the decommissioning cost estimates in Table 1 are listed in Table 12. The figures given may not be totally consistent with each other, as a large part of the labour costs is often hidden in prices of work packages where labour costs are combined with various other costs.

The share of contracted vs. own labour may also be important for costs. In general, the unit man-hour costs for contracted work are higher than those for own labour, but, the difference in total decommissioning costs is harder to assess since the price of contracted work is likely to include risk margins and indirect costs that otherwise would have been covered by contingencies and owner's management costs.

Calculated labour requirements in the estimates in Table 1 are shown in Table 13. Differences are substantial, but less remarkable if they are compared with differences in the assumed scope of work and with differences in estimated waste arisings (Table 10). Just as the waste quantities, the number of man-hours is also assumed substantially higher for Magnox reactors than for the other reactor types, indicating again that a larger total effort is required. The restricted scope — dismantling only to the extent required for termination of the NRC license — is an important explanation among others for the low US estimate.

The large difference between the Finnish estimates is due to several reasons. First, the estimate FI-1 corresponds to a fairly complicated PWR design (VVER-440) — as was already explained in Section 5.4.3. The dismantling of the horizontal steam generators in their narrow space in the containment building is considered particularly difficult and subject to severe working restrictions due to elevated radiation levels. On average the dismantling and disposal costs in FI-1 are estimated at about Mk 112 per kilogram of radioactive waste produced whereas for the FI-2 the average is about Mk 70 per kilogram.

---

1. This range of variation is based on the application of normal currency exchange rates and is not indicative of the actual differences in purchasing powers of the wages.

However, the main difference in labour requirements arises from the effort required by the contaminated systems. The volume occupied by the auxiliary systems, e.g. the water and waste treatment systems, is many times larger in the case of the VVER plant than for the BWR plant. The difference in dismantling effort is further increased by the difference in decommissioning strategies: FI-1 assumes immediate dismantling whereas a 30 year delay is assumed in FI-2. After the delay a considerable part of the equipment that was originally slightly contaminated is assumed to be free for unrestricted release and is not taken into account in the decommissioning cost estimate.

A part of the difference in the Finnish manpower estimates corresponds to different staffing policies. In the case of FI-1 the owner's decommissioning organisation is formed from the plant operating personnel and kept almost constant throughout the whole project period, despite temporary reductions in the extent of actual decommissioning activities. The schedule of the decommissioning activities is affected by the fact that the last spent fuel batches need to be cooled in the reactor ponds for a certain time before they are shipped off-site. In particular this will cause a delay before the dismantling of the second plant unit can start. In the case of FI-2 only the staff strictly required by the decommissioning project is taken into account in the estimate.

The extent to which overhead and administration staff is included in the estimated labour requirements is a possible source of inconsistency among the estimates considered. As was mentioned above, some estimates treat the corresponding costs indirectly through cost factors and, hence, do not take these staff into account in the labour requirements estimate.

Another factor deserving attention in comparison of labour requirements is the productivity of labour. The amount of manpower needed depends on the working methods and the equipment that are planned to be used. The eventual choice of techniques is likely to be determined by the available options at the time of decommissioning, but the present plans are likely to reflect the present situation, and, hence, the differences in existing infrastructures probably affect the estimates.

5.6.2 Materials and equipment

Besides the fact that different working techniques may be used in different decommissioning plans, the prices of the materials and equipment may turn out different in different countries, depending, e.g. on whether they are purchased or leased from abroad, or whether domestic supply of these items can be assumed. In particular, the assumptions concerning the future prices may be different as far as assumptions concerning relative price escalation or technical evolution are concerned.

5.6.3 Costs of waste management and disposal

In many countries no repositories exist yet for radioactive decommissioning waste and the estimates have to be based on assumptions. Table 14 shows the assumptions of the estimates in Table 1. In general, the costs of disposal depend on the techniques used (geologic disposal vs. surface or shallow land disposal) and the scale of operations. A factor of great importance is whether only marginal costs have to be paid — in the case where the wastes are disposed of in an existing repository or an extension of such — or whether the full site preparation, construction and operation costs have to be covered.

The impact of the disposal method was already discussed under Section 5.5 where a reference was made to a study done by the CEC. The CEC carried out under contract with the French (granite), German (salt) and Belgian (clay) relevant bodies an extensive study on the "Cost and mode of financing of geological disposal" [21] during the early 1980s. The huge amount of data and results made available by this study was combined with a survey conducted in 1988 among the agencies in charge of radioactive waste all over Europe, to provide an overall assessment of waste management cost in several European countries. For that purpose, the various economic data were analysed, and, as far as possible, homogenised [18]. The study showed that the economies of scale were significant at low capacities (less than 500 000 m³ for surface disposal, less than 50 000 m³ for deep burial). The economic benefit that might result from the scaling effect at larger capacities appears to be small. Despite the diversity of geological formations and disposal concepts for which economic data were compared, a certain coherence can be detected; thus, for the disposal of low-level waste, costs evolve as a function of site capacity from 2 000 to 6 000 ECU/m³ for deep disposal, and from 1 000 to 3 000 ECU/m³ for near-surface disposal. For deep disposal of medium- and long-lived waste, costs vary as a function of site capacity from 10 000 to 70 000 ECU/m³, and from 400 000 to 1 400 000 ECU/m³ for vitrified high-level waste.

Some data for the estimates of Table 1 are given in Table 14. The estimates from Finland, USA, and Sweden correspond to marginal costs of extending the use of a repository that is assumed already to exist at the time of decommissioning. The case is similar for the low-level wastes of estimate UK-1. However, the other UK estimates are based on a new facility and are, therefore, higher on average.

The total waste management costs depend also on conditioning, packaging and transportation costs as well as on the costs of possible pretreatment to diminish the waste volumes. In the case where the unit disposal costs are high, volume compaction techniques may efficiently reduce the total waste costs.

The optimization of total waste management costs may lead to different technical solutions depending on whether the plant owner is responsible himself for the construction and operating of the repository or whether the wastes will be taken care of by a government facility with a fixed price for the service. A large centralized facility may benefit from economies of scale since the impact of increased transportation distance is likely to be small.

The difference between the users and owners may be important in another way as well. In the USA, at least, where the responsibility is with the states and the Federal Government, the prices for the waste disposal have been increasing during the last few years. Many utilities are afraid that the cost escalation will continue and, therefore, make provisions for very high disposal costs in their estimates.

5.7 Contingencies

The purpose of the contingency allowances is to account for uncertainties in the cost estimates and to provide against unexpected changes in technical plans that may lead to additional costs. The size of contingency allowances depends on the perceived amount of residual uncertainties or technical risks in relation to the degree of conservatism sought for.

The contingency allowance may, therefore, be widely different for different estimates depending on how much is known on the planning subject, how conservative the original plans and individual base estimates are, and what is the purpose of the estimate. In some cases contingencies are not applied as the uncertainties are sufficiently taken into account in the base estimates.

Among the estimates in Table 1, the lowest value of contingency allowance is 10 per cent while the highest is 70 per cent of the basic cost estimate, i.e. the contingencies can make some 10 to 45 per cent of the final cost estimate (see Table 15). The German, Italian and Japanese estimates do not include specific contingency allowances. In some estimates the contingencies are assessed separately for different cost items while in some estimates a general allowance is made in the total cost estimate.

The degree of provision for future uncertainties may be controlled by regulators or even by the law. In Finland the estimates used as a basis for the funding requirements have to be conservative but they shall be made to reflect the current level of prices and the current regulatory requirements. Risks of truly unexpected problems in plant decommissioning or waste management are covered by securities and are not included in considerations of contingency allowance. Therefore, a relatively low overall contingency has been deemed appropriate in the present detailed estimates.

In the USA the level of contingencies varies over a broad range. Some public utility commissions have disallowed any contingency, some allow 50 per cent; in most estimates the average contingency is about 25 per cent. Diverse considerations have been taken into account. In some cases the major reason for high contingency is the perceived increase in waste burial costs. Otherwise the contingency reflects various considerations of project difficulties and costing uncertainties.

For the BNFL estimate from the UK, 12 independent cost elements were identified for the decommissioning of BNFL's fuel cycle plants and reactors and maximum credible variances were put on them. For example, these variances included additional 100 per cent on waste volume estimates, 25 per cent on additional waste treatment costs and 50 per cent on additional intermediate level waste disposal costs. One of the elements was taken as reactor decommissioning and a maximum credible variance of 70 per cent of additional costs was assigned to it. The variances were considered independent and so were added statistically so as to yield an overall contingency which came out at approximately 40 per cent for the total decommissioning cost estimate. Work is ongoing to more accurately assess and cost the various cost elements of decommissioning which will lead to reductions in the credible variances and hence the contingency level.

Since the contingency allowances reflect very different considerations, there is no single figure that could be suggested to be generally applicable. A common value of 25 per cent was used in the 1986 NEA decommissioning study for comparison purposes, but because of the differences in cost estimation practices and in the level of detailness of the plans, such a normalization may well broaden the inconsistency rather than reduce it.

6. CONCLUSIONS

Considerable attention has been drawn in public discussion to the apparent lack of agreement between the decommissioning cost estimates reported for various nuclear power plants in different countries and regions. The survey made for this study confirms that, indeed, the numbers do look quite different if compared as such. However, the analysis of the background of the estimates shows that, in spite of the apparent discrepancies, a fairly consistent picture can be obtained.

6.1 Decommissioning can take various practical forms

In spite of common approaches, the legal and regulatory frameworks that govern nuclear energy use in OECD countries still remain specific to each country. Several reactor types and plant designs are employed. Different technical solutions to radioactive waste disposal are being planned and investigated. It is, therefore, only natural that a variety of strategies and plans are being studied or implemented for taking care of the shut-down nuclear power plants and fuel cycle facilities.

The word "Decommissioning" broadly means all activities that begin after the facility operations have ceased and intend to bring the site into an environmentally safe condition. Consequently, it may refer to widely different practical solutions which include different technical approaches and different time schedules.

6.2 The size of decommissioning projects will be different

For cost considerations one of the most influential elements is the scope of the plan. Some of the plans cover only the decommissioning activities up to Stage-1 or 2, others up to Stage-3. Among the plans for Stage-3, there are differences in the extent to which it is foreseen to be necessary to dismantle and demolish plant components and structures. In some cases everything will be removed; in some other cases the plan is restricted to dismantling only those parts that are considered radioactive.

The planned scope of the dismantling work has an important effect on the extent of the decommissioning project. However, even in cases where everything will be removed, what actually has to be done at individual plants varies significantly. Reactor type differences are one reason for this variation. Despite having the same basic functions the gas-cooled reactors, heavy-water reactors and light-water reactors all have specific technical characteristics that require different handling at decommissioning. Even for the same reactor type the individual designs may be significantly different; for the same reactor supplier the successive vintages (models) of the same type may also show significant difference.

The estimated waste arisings from decommissioning give an indication of the physical size of structures and the equipment that are planned to be removed. In the case of the examples submitted for this study the estimated amounts of radioactive waste per plant unit show a variation by a factor of five or ten, depending on whether weights or volumes are compared. The requirements for disposal space will be proportional to the conditioned waste volumes. Also the labour effort needed in dismantling and demolition is likely to be larger, the larger the waste amounts are, albeit, the work required to produce a unit amount of waste largely depends on materials and working condition, e.g. the radiation level.

44

6.3 Different cost assumptions have understandable reasons

The unit costs for labour and waste disposal are usually the most important cost factors in decommissioning projects. There are good reasons for variable assumptions concerning the level of these costs in various countries and regions.

A look at ILO statistics of labour rates in late 1980s reveals that the range of variation in hourly wages of industrial workers — when converted to US dollars using normal exchange rates — exceeds a factor of two in the countries considered in this study (Annex 6). This comparison is for the gross wages. Possible differences in other wage-related costs to the employer, e.g. the mandatory contributions to health care and pension scheme, may further extend the range of variation.

Furthermore, decommissioning cost estimates are based on assumed labour costs and, therefore, may not only reflect the current cost situation but also future expected developments. For waste disposal costs this distinction is even more important: as long as the requirements and criteria for radioactive waste disposal are pending final decision, widely different assumptions can be defended.

The judgment of prevailing uncertainties in the unit costs for various decommissioning activities may result in widely different contingency allowances, largely depending on the degree of conservatism that is considered necessary in the estimate. The degree of conservatism is likely to depend on the purpose of the estimate. In the case where the estimate is used as a basis for fund-raising for decommissioning activities and strong assurance is required for the timely availability of adequate funds, the total contingency allowance for future uncertainties may be much higher than in the case where normal costing practices of construction industries are followed (Table 15). However, since the basic cost estimates (without contingency allowances) already reflect varying degrees of conservatism, comparison of contingency allowance alone is just as misleading as is the comparison of the basic cost estimates without contingencies.

In the end, it has to be accepted that engineering judgment plays an important role in all cost estimating, and the estimator's background and experience give rise to different judgments. The discussion in this report has focused on factors that produce material differences to decommissioning projects; nevertheless, this does not imply that the impact of judgmental differences should be played down.

6.4 Exchange rate changes may lead to fallacious conclusions

International cost comparisons are meaningful only to the extent that the currency conversions made are representative of actual cost parities between countries. A simple consideration of the decommissioning cost data in the 1986 NEA report suffices to raise considerable doubts concerning the use of exchange rates for this purpose.

The cost data in the 1986 report were given in January 1984 US dollars. Table 16 shows the corresponding exchange rates together with the exchange rates as of January 1990. It is seen that the values of some national currencies against the US dollar have increased by more than 60 per cent whereas in two cases the change is only around ten per cent. In Table 17 the original decommissioning cost estimates of the 1986 report (immediate dismantling strategy) are first escalated to correspond to January 1990 price levels and then converted to US dollars using the exchange rates as of January 1990. The comparison looks quite different from the 1986 report in the sense that the US estimates now seem considerably lower than the other cost estimates.

The comparisons among European cost estimates are not affected dramatically — although in absolute terms the dollar-based cost estimates now look totally different. Indeed, Table 16 confirms that, with the exception of the UK, the changes in exchange rates against the US dollar have been similar for most European currencies and also for the Japanese yen. In recent years the bias caused by the use of exchange rates is likely to be most severe for the comparison of US and Canadian estimates with estimates from other OECD countries.

The problem of fluctuating exchange rates has been widely recognized in many contexts, but there are no simple, universally applicable means to circumvent the problem. Therefore, international comparisons should be considered with caution, recognizing that the results would probably turn out somewhat differently if some other basis had been chosen for the monetary conversions.

6.5 Consistency despite discrepancies

The basic conclusion of the study is that there is no reason to expect that the decommissioning costs should be similar or even nearly similar for all nuclear power plants in the world. The fundamental cause is that, for several technical or institutional reasons, the extent of physical work that will be done is quite different at different facilities. If the comparison is focused on cost estimates (not on costs), the variations may well be larger still because many important regulatory and policy issues so far remain unsettled. Further divergence is caused by different cost conditions prevailing or projected in different countries.

A different picture is obtained if the estimated decommissioning costs are compared with the corresponding quantities of waste arisings. The estimated amount of wastes can be taken as crudely descriptive of the physical size of the project. Before doing the comparison, however, some of the cost estimates have to be slightly adjusted to allow for significant differences in their coverage.

— Firstly, since some of the estimates are restricted to the costs arising from dismantling the radioactive parts only, the comparison is constrained to the costs attributable to these parts. Correspondingly, only radioactive waste amounts will be considered. In cases where the original cost estimate was for complete dismantling and site restoration, it is assumed that 80 per cent of the estimated costs are attributed to the radioactive parts.

— Secondly, those estimates that originally did not include the costs for waste handling and disposal, have been increased by 20 per cent to make them more comparable with other estimates.

— Thirdly, the British estimates for gas-cooled reactors have been slightly adjusted to remove the cost for defuelling. In other cost estimates, the defuelling costs are not included or do not contribute so much. The original British estimates (for GCRs) have been reduced by 10 per cent.

Table 18 shows the modified cost estimates as proportioned to the estimated weights of the radioactive waste quantities produced. The resulting costs per unit waste quantity still show a range of variation by a factor of about 2.3, but this is significantly less than the factor of 5.3 which is obtained if the estimates are compared on the basis of costs per plant unit (the first column from the right of the table). A variation of this size is not surprising: a similar range of variation is seen in estimates for plant construction costs [2].

If the comparison is restricted to light-water reactors only, the calculated costs per unit of radioactive waste are surprisingly close to each other in all cases but one. If the estimate for Loviisa power station (FI-1) is excluded, the estimates for LWRs are within ±35 per cent from the average. The Loviisa station represents a Soviet VVER-440 which is significantly different from the other PWR designs used in OECD countries.

In fact, if the comparisons are restricted to reasonably similar facilities, the variation in the costs (before any adjustments) per plant unit can also be reduced. Hence, for instance, the range of variation in costs per plant unit in Table 18 is by a factor of 2.3 for BWRs, and for most PWRs the variation is only by a factor of 1.8 (FI-1 and UK-4 fall outside this range).

46

It should be noted that the numerical values described here should not be taken too literally, since the number of the estimates submitted for this study is limited and the monetary conversion is affected by fluctuating exchange rates.

The comparisons in Table 18 show that despite apparent discrepancies, a considerable degree of consistency in the estimates is revealed as soon as a reasonable basis of comparison is used. The cost estimates have to be considered in relation to the actual work required in each individual case. Comparing the decommissioning cost estimates solely on the basis of costs per unit power capacity is misleading. Although the dimensions of the main reactor components are likely to be larger for the higher power capacities, the power capacity alone poorly reflects the extent of work required by various decommissioning projects.

6.6 Decommissioning costs can be estimated with reasonable accuracy

The present cost estimates for future decommissioning projects are based on the technical experience and costing expertise developed in various related projects in nuclear industries. Considerable experience already exists on the decommissioning of research, demonstration and smaller prototype reactors, and several on-going or planned projects will extend the experience to larger reactors and fuel cycle facilities. Most techniques required for decommissioning are also being used in maintenance and repair work at operating nuclear facilities.

Consequently, there is no uncertainty about the availability of feasible techniques for decommissioning. Further technical development will probably bring in new methods, but the present plans and cost estimates can be wholly based on technical means which are available now. Normal costing uncertainties naturally remain, and sometimes these are amplified by open policy issues, such as those concerning the waste disposal practices. However, there is also information from a number of completed decommissiong projects which give evidence that decommissioning costs can be estimated with reasonable accuracy.

REFERENCES

[1]   OECD/NEA (Nuclear Energy Agency) (1986), *Decommissioning of Nuclear Facilities: Feasibility, Needs and Costs*, Paris.

[2]   OECD/NEA-IEA (Nuclear Energy Agency-International Energy Agency) (1989), *Projected Costs of Generating Electricity from Power Stations for Commissioning in the Period 1995-2000*, Paris.

[3]   Special report, *Outlook on Decommissioning Costs*, Nucleonics Week, Sept. 27, Nuclear Fuel, Oct.1, Inside N.R.C., Oct. 8, 1990.

[4]   IAEA (International Atomic Energy Agency) (1986), Technical Reports Series No. 267, *Methodology and Technology of Decommissioning Nuclear Facilities*, Vienna.

[5]   OECD/NEA (Nuclear Energy Agency), *OECD/NEA Co-operative Programme on Decommissioning, Achievements and Perspectives 1985-1990* (in preparation).

[6]   IAEA (International Atomic Energy Agency) (1983), Technical Reports Series No. 230, *Decommissioning of Nuclear Facilities: Decontamination, Disassembly and Waste Management*, Vienna.

[7]   AIF (Atomic Industrial Forum, Inc.) (1986), AIF/NESP-036, *Guidelines for Producing Commercial Nuclear Power Plant Decommissioning Cost Estimates*, Bethesda.

[8]   OECD/NEA (Nuclear Energy Agency) (1984), *Long-Term Radiological Aspects of Management of Wastes from Uranium Mining and Milling*, Paris.

[9]   ICRP (International Commission on Radiological Protection) (1985), ICRP publication No. 46, *Radiation Protection Principles for the Disposal of Solid Radioactive Waste*, Pergamon Press, Oxford.

[10] IAEA (International Atomic Energy Agency) (1988), Safety Series No. 89, *Principles for the Exemption of Radiation Sources and Practices from Regulatory Control*, Vienna.

[11] CEC (Commission of the European Communities) (1988a), Radiation Protection No.43, *Radiological Protection Criteria for the Recycling of Materials from the Dismantling of Nuclear Installations*, Luxembourg.

[12] IAEA (International Atomic Energy Agency) (1985), Safety Series No. 71, *Acceptance Criteria for Disposal of Radioactive Wastes in Shallow Ground and Rock Cavities*, Vienna.

[13] BURHOLT, G.D. and MARTIN, A. (1988), *The Regulatory Framework for Storage and Disposal of Radioactive Waste in the Member States of the European Community*, EUR 11292, CEC, Brussels.

[14] LAGUARDIA, Thomas S., PE (1987), *Decommissioning Cost Estimating and Contingency Application*, Proceedings of the 1987 International Decommissioning Symposium (Pittsburgh, Pennsylvania, October 4-8, 1987), edited by Gail A. Tarcza, CONF-871018-Vol.2 (DE87012822), Westinghouse Hanford Company, Richland, Washington.

[15] SKB (Svensk Kärnbränslehantering AB — Swedish Nuclear Fuel and Waste Management Co.) (1986), Technical Report 86-18, *Technology and Costs for Decommissioning the Swedish Nuclear Power Plants*, Stockholm.

[16] BERNOW, Stephen and BIEWALD, Bruce (1987), *Nuclear Power Plant Decommissioning: Cost Estimation for Planning And Rate Making*, Public Utilities Fortnightly, October 29.

[17] DUBOURG, M. (1990), *Design Features Adopted to Facilitate Decommissioning*, Decommissioning of Nuclear Installations, EUR 12690, CEC, Brussels.

[18] ZACCAI, H. (1990), *Evaluation of Storage and Disposal Costs for Conditioned Radioactive Waste in Several European Countries*, EUR 12871, CEC, Brussels.

[19] WOOD, Janet (1990), *Cost Lessons Learnt from Decommissioning Shippingport*, Nuclear Engineering International, September.

[20] ILO (International Labour Organisation) (1990), *Year Book of Labour Statistics 1989-1990*, Geneva.

[21] CEC (Commission of the European Communities) (1988b), *Coûts et modes de financement de l'évacuation géologique des déchets radioactifs*, EUR 11837, Brussels.

Table 1. Recent decommissioning cost estimates for commercial nuclear power plants

| REACTOR TYPE | COUNTRY | ESTIMATE CODE | FACILITY DESCRIPTION | | | | | MODE OF DECOMMISSIONING | ESTIMATED DECOMMISSIONING COSTS (2) | | YEAR OF ORIGINAL ESTIMATE | CPI ESCALATOR APPLIED (4) | EXCHANGE RATE AS OF JANUARY 1990 (NCUs per USD) |
			Capacity (1) (MWe)	Start of construction	Start of commercial operation	Assumed operating lifetime (years)	NSSS* SUPPLIER		as given in the response to the questionnaire (3)	in millions of US dollars of January 1990			
PWR (5)	FINLAND	FI-1	2 x 465	1970, 71	1977, 78	30	AEE	STAGE 3	882 M FIM (1989)	237	1987	1.076	4.003
PWR (6)	GERMANY	GER-1	1204	1970	1974	40	KWU	STAGE 3	346 M DM (1985)	218	1985	1.067	1.692
PWR (6)	GERMANY	GER-2	1204	1970	1974	40	KWU	30 years + STAGE 3	325 M DM (1985)	205	1985	1.067	1.692
PWR	JAPAN	J-1	1160	1980s	1980s	40	MHI	STAGE 1 +(5-10 years) + STAGE 3	30.2 b yen (1984)	225	1984	1.084	145.2
PWR	SWEDEN	SW-2	920	1968	1975	40	WH	STAGE 3	805 M SEK (1990)	130	1986	1.000	6.171
PWR	UK	UK-4	1155	1988	1995	40	PPC	STAGE 3	253 M £ (1990)	418	1990	1.000	0.605
PWR (6)	US	US-1	1175	1960s	1960s	40	WH	STAGE 3	103.5 M USD (1986)	120	1986	1.162	1.000
BWR (7)	FINLAND	FI-2	2 x 735	1974, 75	1979, 81	40	ASEA	STAGE 1 + (30 years) + STAGE 3	793 M FIM (1990)	198	1989	1.000	4.003
BWR (8)	ITALY	ITA	160	1959	1964	18	GE	STAGE 1	65000 M Lira (1989)	54.8	1989	1.064	1262
BWR	JAPAN	J-2	1100	1980s	1980s	40	TO or HIT	STAGE 1 + (5-10 years) + STAGE 3	31.4 b yen (1984)	234	1984	1.084	145.2
BWR	SWEDEN	SW-1	780	1968	1976	40	ASEA	STAGE 3	940 M SEK (1990)	152	1986	1.000	6.171
GCR	SPAIN	SPA	500	1962	1972	17	CEA	STAGE 1 + STAGE 2 +(25 years) + STAGE 3	45 b peseta (1990)	410	1990	1.000	109.68
GCR (9)	UK	UK-1	8 x 60	~1954	~1956	40	UKAEA	STAGE 1 + STAGE 2 +(60-90 years)+STAGE 3	836 M £ (1989)	1488	1988	1.077	0.605
GCR (10)	UK	UK-2	2 x 219	1957-63	1962-72	30	APC, BNDC NNC, TNPG	STAGE 1 + STAGE 2 +(90 years) + STAGE 3	634 M £ (1990)	1048	1990	1.000	0.605
AGR	UK	UK-3	2 x 660	1966-80	1977-88	25	APC, BNDC NNC, TNPG	STAGE 1 + STAGE 2 +(90 years) + STAGE 3	601 M £ (1990)	993	1990	1.000	0.605
HWR	CANADA	CAN	600	1973	1979	30	AECL	STAGE 1 + (32 years) + STAGE 3	264 M CAD (1989)	238	1989	1.055	1.172

(1) Design gross capacity.
(2) Costs for all facilities in the fourth column.
(3) The number in parenthesis is the base year for the money value.
(4) Calculated as x/y   x = the national consumer price index (CPI) for January 1990,
    y = the national consumer price index (CPI) for January of the base year of the money value.

* NSSS: Nuclear Steam Supply System.

(5) Basic design in Soviet VVER-440. Equipped with a Westinghouse type containment with ice condenser.
(6) With cooling tower.
(7) Equipped with internal main recirculation pump.
(8) Dual cycle.
(9) Each plant has 4 external boilers and cooling tower.
(10) Average of 8 stations. First 6 have steel pressure vessels with external boilers. Last 2 have concrete pressure vessels with internal boilers.

AECL : Atomic Energy of Canada Ltd. (Canada)
AEE : Atomenergoexport (USSR)
APC : Atomic Power Company (UK)
ASEA : ASEA-ATOM (Sweden)
BNDC : British Nuclear Design and Construction (UK)
CEA : Commissariat à l'Energie Atomique (France)

GE : General Electric (US)
HIT : Hitachi Ltd. (Japan)
KWU : KraftWerk Union (Germany)
MHI : Mitsubishi Heavy Industry (Japan)
NNC : National Nuclear Corporation (UK)

PPC : PWR Power Co. Ltd. (UK)
TNPG : The Nuclear Power Company (UK)
TO : Toshiba Corp. (Japan)
UKAEA : United Kingdom Atomic Energy Authority (UK)
WH : Westinghouse Electric Corp. (US)

Table 2. **Decommissioning cost estimates for Spanish nuclear power plants**

PLANT	FACILITY DESCRIPTION				ESTIMATED DECOMMISSIONING COSTS (millions USD of January 1990)
	Type	Gross Capacity (MWe)	Start of construction	Start of commercial operation	
José Cabrera	PWR	160	1959	1968	90
S.Ma Garona	BWR	460	1962	1971	160
Vandellos I (a)	GCR	500	1962	1972	410
Vandellos II	PWR	992	1979	1988	220
Almaraz I-II	PWR	2 x 930	1973/74	1981/83	400
Asco I, II	PWR	2 x 930	1972/72	1983/85	400
Cofrentes	BWR	900	1974	1984	300
Trillo I	PWR	1043	1980	1988	230

(a)   Detailed estimate has been made for this plant. The estimate is also included in Table 1.

Table 3. **Decommissioning cost estimates for Swedish nuclear power plants**

PLANT	FACILITY DESCRIPTION				ESTIMATED DECOMMISSIONING COSTS (million SEK of 1986)		
	Type	Gross capacity (MWe)	Start of construc-tion	Start of commercial operation	Total decommis-sioning	Shutdown operation	Transport and final disposal of waste
Oskarsham 1	BWR	462	1966	1972	410	}	}
2	BWR	630	1969	1975	470	} 190	} 150
3	BWR	1200	1980	1985	750	}	}
Barsebäck 1	BWR	615	1971	1975	460	}	}
2	BWR	615	1973	1977	490	} 110	} 90
Ringhals 1 (a)	BWR	780	1969	1976	540	}	}
2 (a)	PWR	920	1970	1975	460	}	}
3	PWR	960	1972	1981	460	} 310	} 190
4	PWR	960	1973	1983	460	}	}
Forsmark 1	BWR	1005	1973	1980	670	}	}
2	BWR	1005	1975	1981	660	} 190	} 170
3	BWR	1150	1979	1985	760	}	}

(a)   Detailed estimates were prepared for these plants. These estimates are also included in Table 1.

Table 4. Estimated decommissioning costs in NEA studies
(Plant type and size are shown in parentheses)

COUNTRY	PRESENT STUDY NCUs (Base year shown in [ ])	1986 NEA REPORT ON DECOMMISSIONING NCUs, January 1984	1989 NEA/IEA REPORT ON ELECTRICITY GENERATION COSTS NCUs, January 1987
Belgium (1)	(d)	(d)	8.9 (PWR 1390 MW)
Canada (2)	264 (HWR 600 MW)[1989](b)	263 (HWR 4x515 MW)(a) 213 (HWR 4x515 MW)(b)	184 (HWR 881 MW) 70 (HWR 400 MW)
Finland (3)	882(PWR 2x445MW)[1989](a) 793(BWR 2x710MW)[1990](b)	430 (PWR 2x440 MW)(a) 767 (BWR 2x710 MW)(b)	696 (PWR 1000 MW)
France (4)	(d)	(d)	1.4 (PWR 1390 MW)
Germany (5)	346 (PWR 1204 MW)[1985](a) 325 (PWR 1204 MW)[1985](b)	283 (PWR 1200 MW)(a) 291 (PWR 1200 MW)(b) 313 (BWR 800 MW)(a) 329 (BWR 800 MW)(b)	525 (PWR 1256 MW)
Italy (6)	65 (BWR 160 MW)[1989](h)	(d)	471 (PWR 945 MW)
Japan (7)	30 (PWR 1160 MW)[1984](e) 31 (BWR 1100 MW)[1984](e)	(d)	30 (LWR 1100 MW)
Netherlands (8)	(d)	(d)	701 (PWR 1300 MW)
Spain (9)	45 (GCR 500 MW) [1990]	(d)	26 (PWR 950 MW)
Sweden (10)	940 (BWR 780 MW)[1990](a) 805 (PWR 920 MW)[1990](a)	440 (BWR 440 MW)(a) 560 (BWR 570 MW)(a) 640 (BWR 750 MW)(a) 840 (BWR 900 MW)(a) 920 (BWR 1050 MW)(a) 640 (PWR 880* MW)(a)	(d)
United Kingdom (11)	836(GCR 8x60 MW)[1989](f) 634(GCR 2x220 MW)[1990](g) 601(AGR 2x660 MW)[1990](g) 253 (PWR 1155 MW)[1990](a)	(d)	233 (PWR 1175 MW)
United States (12)	104 (PWR 1100 MW)[1986](a)	88 (PWR 1175 MW)(a) 109 (PWR 1175 MW)(b) 143 (PWR 1175 MW)(c) 100 (BWR 1155 MW)(a) 125 (BWR 1155 MW)(b) 165 (BWR 1155 MW)(c)	114 (PWR 1144 MW)

NCU: National Currency Unit.
- (1) Billion Belgian Franc.
- (2) Million Canadian Dollar.
- (3) Million Markka.
- (4) Billion French Franc.
- (5) Million DM.
- (6) Billion Lira.
- (7) Billion Yen.
- (8) Million Gulden.
- (9) Billion Peseta.
- (10) Million Krona.

- (11) Million Pound.
- (12) Million US dollar.

- (a) Stage-3, immediately.
- (b) Stage-1, ~30 years storage, Stage-3.
- (c) Stage-2, 100 years storage, Stage-3.
- (d) Not included.
- (e) Stage-1, 5-10 years storage, Stage-3.
- (f) Stage-1, Stage-2, 60-100 years storage, Stage-3.
- (g) Stage-1, Stage-2, 90 years storage, Stage-3.
- (h) Stage-1.

* The capacity is changed from 915 MWe (in the 1986 NEA report) to 880 MWe on the basis of suggestions from the Swedish expert.

Table 5. Comparison of large components in Shippingport with those of other nuclear power plants (a)

Component	Shippingport (USA) 72 MWe PWR	Trojan (USA) 1175 MWe PWR	Ringhals 2 (Sweden) 800 MWe PWR	Ringhals 1 (Sweden) 750 MWe BWR
Reactor Pressure Vessel: OD/L (m) Weight incl. head (t)	3.0/9.4 (b) 248 (b)	4.6/12.6 350	4.4/13.0 209	6.3/21.5 530
Steam Generator: Weight total (t)	100	312	296	NA
Main circ. pump: Weight (t)	17	85	35	NA
Pressurizer: Weight (t)	40	89	86	NA

(a) Quoted from "Waste from Decommissioning of Nuclear Power Plants", 1990, Scandpower, Norway.
(b) Excluding the neutron shield tank; weight for the RPV including the neutron shielding tank and internals prepared for shipment: 700 tons.

Table 6. Free release criteria for radioactive waste assumed in the estimates of Table 1

Estimate	Assumed Free Release Criteria
CAN	Not assumed
FI-1 FI-2	Not assumed 10 Bq/g
GER-1 GER-2	{3.7 Bq/g 0.37 Bq/cm² : beta-gamma {0.37 Bq/g 0.037 Bq/cm² : alpha
ITA	Not assumed
J-1 J-2	3.7×10^6 Bq/t 3.7×10^6 Bq/t
SPA	Not assumed
SW-1 SW-2	3×10^5 Bq/t 3×10^5 Bq/t
UK-1 UK-2 UK-3 UK-4	0.4 Bq/g 0.4 Bq/g 0.4 Bq/g 0.4 Bq/g
US-1	Not assumed (a)

(a) Release criteria for radioactive wastes are based on surface contamination limits in Regulatory Guide 1.86, Table 1, and site specific pathways analyses for the calculation of "disposition criteria" which were based upon an annual dose limit of 1 to 25 mrem per year.

Table 7. **Radioactive waste disposal facilities assumed in the estimates of Table 1**

ESTIMATE	WASTE	ASSUMED FACILITIES	CURRENT SITUATION
CAN	ILW (a) & LLW (b)	Off-site facility	Not decided
FI-1	ILW & LLW	Underground repository to be built at the site	Commissioning expected in 1997-2000
FI-2	ILW & LLW	Underground repository to be built at the site	Under construction for reactor operations wastes (start-up in 1992); to be enlarged later for decommissioning wastes
GER-1 } GER-2 }	ILW & LLW	Deep geological formation	Not decided
ITA	(c)	-	-
J-1 } J-2 }	(c)	-	-
SPA	(c)	-	-
SW-1 } SW-2 }	ILW & LLW	Underground repository at the Forsmark site	Exists for operational waste Can be enlarged when the need arises
UK-1	ILW	Nirex deep land repository	Attempting to get construction permission
	LLW	Drigg shallow land repository	Exists
UK-2 } } 3 } 4 }	Low-level operational waste	Drigg shallow land repository	Exists
	ILW & LLW	Nirex deep land repository	Attempting to get construction permission
US-1	LLW & ILW	Off-site facility	Not decided

(a)   Intermediate Level Radioactive Waste.
(b)   Low Level Radioactive Waste.
(c)   Disposal costs are excluded in the estimate.

Table 8. Scopes of estimates in Table 1

ESTIMATE	SCOPE OF PROJECT	INCLUSION OF COST COMPONENTS (a)	
		Defuelling (b)	Radioactive waste disposal
CAN (c)	Greenfield	Yes	Yes
FI-1 & 2	Radioactive part only	No	Yes
GER-1 & 2	Greenfield	No	Yes
ITA	(Stage-1)	No	No
J-1 & 2	Greenfield	No	No
SPA	Greenfield	No	No
SW-1 & 2	Greenfield	Yes	Yes
UK-1, 2, 3, 4	Greenfield	Yes	Yes
US-1	Radioactive part only	No	Yes

(a) Yes = includes   No = excludes

(b) Back-end fuel cycle costs are not included in any estimates.

(c) The Canadian estimate includes costs for:
   — removal and disposal of replacement pressure tubes,
   — disposal of operational wastes,
   — demolition and disposal of the spent fuel interim dry storage facility.

Table 9. Comparison of radioactive waste amounts from decommissioning of a Magnox station and a PWR

	Magnox (Berkeley, 2 x 138 MW)	PWR (Sizewell "B" 1200 MW single unit)
Concrete	24 000 t	8 000 t
Steel	19 000 t	4 500 t
Graphite	4 000 t	-
TOTAL	47 000 t	12 500 t

Table 10. Estimated amounts of radioactive waste

ESTIMATE	REFERENCE PLANT(s) (A) (reactor type, number of units and capacity)	AMOUNTS OF RADIOACTIVE WASTE FROM DECOMMISSIONING ACTIVITIES* (B) (ton)	VOLUMES OF RADIOACTIVE WASTE TO BE DISPOSED OF (m³)	WASTE AMOUNTS PER UNIT CAPACITY (B/A) (ton/MWe)
CAN	HWR 1 x 600 MWe	8 342 (a)	17 500	13.9 (a)
FI-1 FI-2	PWR 2 x 465 MWe BWR 2 x 735 MWe	7 700 11 200	13 000 29 300	8.3 7.6
GER-1 GER-2	PWR 1 x 1204 MWe PWR 1 x 1204 MWe	10 300 10 300	3 300 (b) 2 500 (b)	8.6 8.6
ITA	BWR 1 x 160 MWe	-	-	-
J-1 J-2	PWR 1 x 1160 MWe BWR 1 x 1100 MWe	11 800 11 300	- -	10.2 10.3
SPA	GCR 1 x 500 MWe	14 250	17 000	28.5
SW-1 SW-2	BWR 1 x 780 MWe PWR 1 x 920 MWe	5 540 4 065	10 000 7 500	7.1 4.4
UK-1 UK-2 UK-3 UK-4	GCR 8 x 60 MWe GCR 2 x 219 MWe AGR 2 x 660 MWe PWR 1 x 1155 MWe	81 372 37 786 23 800 12 500	125 146 52 000 36 619 -	169.5 86.3 18.0 10.8
US-1	PWR 1 x 1175 MWe	8 500 (a)	17 830	7.2 (a)

*   Before any treatment.

(a)   Volume of radioactive waste to be disposed of was provided in terms of cubic-meters including packaging. The weight of raw radioactive waste was calculated by the Secretariat using the ratio between the raw weight and volume to be disposed of, which is obtained from information in FI-1, FI-2, SW-1 and SW-2.

(b)   Thorough volume reduction was applied.

Table 11. Decommissioning costs in proportion to power capacities and radioactive waste amounts

ESTIMATE	REACTOR TYPE	DECOMMISSIONING COST/CAPACITY (a) ($/We)	DECOMMISSIONING COST/WASTE AMOUNT (b) ($/kg)
CAN	HWR	0.40	28.5 (c)
FI-1	PWR	0.25	30.8
FI-2	BWR	0.13	17.7
GER-1	PWR	0.18	21.2
GER-2	PWR	0.17	19.9
ITA	BWR	0.34	- (d)
J-1	PWR	0.19	19.1
J-2	BWR	0.21	20.7
SPA	GCR	0.82	28.8
SW-1	BWR	0.19	27.4
SW-2	PWR	0.14	32.0
UK-1	GCR	3.10	18.3
UK-2	GCR	2.39	27.7
UK-3	AGR	0.75	41.7
UK-4	PWR	0.36	33.4
US-1	PWR	0.10	14.1 (c)

(a) Design gross capacity.

(b) Weight of radioactive waste before any treatment.

(c) Weight of radioactive waste calculated by the Secretariat from the volume of radioactive waste to be disposed of.

(d) The radioactive waste data is not available. The plant will be decommissioned up to Stage-1.

Table 12. Labour cost assumptions in the estimates of Table 1

ESTIMATE	DESCRIPTION	LABOUR COSTS* ASSUMED IN ESTIMATE (NCU**)	ADJUSTED VALUE*** (US$ OF JANUARY 1990/HOUR)
CAN	Direct labour Project management, engineering, etc.	26 CAD/hr 56 CAD/hr	23.4 $/hr 50.4 $/hr
FI-1	Contracted staff Own staff	115-270 FIM/hr 75-168 FIM/hr	34.2-80.2 $/hr 22.3-49.9 $/hr
FI-2	Contracted staff Own staff	129-315 FIM/hr 100-168 FIM/hr	34.7-84.7 $/hr 26.9-45.2 $/hr
GER 1 & 2	N.A.		
ITA	Average	70 000 LIRA/hr	59.0 $/hr
J-1 & 2	N.A.		
SPA	N.A.		
SW	Contracted staff Own staff	200-500 SEK/hr 200-300 SEK/hr	40.5-101.4 $/hr 40.5- 60.8 $/hr
UK-1	Average	38 000 £/year (a)	43.1 $/hr
UK-2, 3 & 4	Worker Supervising staff	12.5 £/hr 22.0 £/hr	20.7 $/hr 36.4 $/hr
US-1	Average	35.36 $/hr (b)	41.1 $/hr

* Total labour costs which include not only wages/salaries before tax deduction but also employer's mandatory contributions to health care, insurance and pension schemes.

** National currency unit in the year of estimate.

*** Adjusted by the Secretariat.

(a) 1 man-year = 1570 man-hours.
(b) Average total labour cost.

Table 13. **Estimated labour requirements in the estimates of Table 1** (a)

ESTIMATE	TOTAL LABOUR REQUIREMENT (man-years)
CAN	1 115
FI-1	2 920
FI-2	1 560
GER-1	-
GER-2	-
ITA	306 (b) (550 000)
J-1	-
J-2	-
SPA	-
SW-1	~1 000
SW-2	~1 000
UK-1	5 128
UK-2	5 732 (b) (~9 000 000)
UK-3	5 732 (b) (~9 000 000)
UK-4	-
US-1	301 (b) (627 000)

(a)   Each data corresponds to all facilities covered in each estimate.

(b)   Original data was provided in terms of man-hour and is put in parentheses. The value in terms of man-year was converted from the original data using following conversion factors:

ITA	:	1800 man-hours/man-year
UK-1	:	1570 man-hours/man-year
UK-2,3,4	:	1570 man-hours/man-year
US-1	:	2080 man-hours/man-year

Table 14. Estimated unit cost for radioactive waste disposal in the estimates of Table 1

(in US dollars of January 1990)

ESTIMATE	WASTE	PACKAGING (per m³)	TRANSPORTATION (per m³)	DISPOSAL (per m³)	TOTAL (per m³)
CAN	Category 1 (a)	-	} 2 250 /trip (b)	603	-
	Category 2 (a)	-		603	-
	Category 3 (a)	-		3 871	-
FI-1	Average	-	-	1 633	-
FI-2	Average	336	65	1 371	1 772
GER-1	Average	-	-	-	19 874
GER-2	Average	-	-	-	17 405
ITA	-	-	-	-	-
J-1	-	-	-	-	-
J-2	-	-	-	-	-
SPA	Average	1 500	400	2 000	3 900
SW-1	Average	-	-	1 014	-
SW-2	-	-	-	1 014	-
UK-1	ILW (c)	1 780	712	2 314	4 806
	LLW (c)	-	-	-	1 157
UK-2	ILW (c)	-	-	-	1 240-7 851
	LLW (c)	-	-	-	826-2 479
UK-3	ILW (c)	-	-	-	1 240-7 851
	LLW (c)	-	-	-	826-2 479
UK-4	-	-	-	-	-
US-1	Activated	-	1 612	6 289	-
	Contaminated	-	883	1 664	-
	Other	-	1 133	3 897	-

(a) Category 1 : < 0.5 mr/hr
   Category 2 : > 0.5 mr/hr, < 5 mr/hr
   Category 3 : > 5 mr/hr

(b) "1 trip" approximately correspond to 25 tons of packaged waste.

(c) ILW (Intermediate Level Waste) : < 4 G Bq/t alpha and < 12 G Bq/t beta-gamma.
   HLW (High Level Waste) : > 4 G Bq/t alpha or > 12 G Bq/t beta-gamma.

Table 15. Contingency assumptions for Table 1 cost estimates

ESTIMATE	CONTINGENCY ASSUMPTION
CAN	17 %
FI-1	10 %
FI-2	15 % *
GER-1	Not applied
GER-2	Not applied
ITA	Not applied
J-1	Not applied
J-2	Not applied
SPA	15 %
SW-1	20 %
SW-2	20 %
UK-1	70 %
UK-2	33 %
UK-3	33 %
UK-4	50 %
US-1	25 %

* Plus about 7 % for uncertainty in applicable exemption limits.

Table 16. Exchange rates for selected OECD countries

COUNTRY	JANUARY 1984 NCUs (a) per US$	JANUARY 1990 NCUs (a) per US$	CHANGE IN VALUE OF NCU AGAINST US$
BELGIUM	55.64	35.47	57 %
CANADA	1.24	1.17	6 %
FINLAND	5.81	4.003	45 %
FRANCE	8.35	5.76	45 %
GERMANY	2.72	1.69	61 %
ITALY	1 659.50	1 262.00	31 %
JAPAN	232.20	145.20	60 %
SPAIN	156.70	109.68	43 %
SWEDEN	8.00	6.17	30 %
UNITED KINGDOM	0.69	0.61	11 %
UNITED STATES	1.00	1.00	-

(a)   NCU stands for national currency unit.

Table 17. Cost estimates (immediate Stage-3) in 1986 NEA report presented in US dollar of January 1984 and 1990

COUNTRY	ORIGINAL ESTIMATES* IN 1986 NEA REPORT MUSD, January 1984	ADJUSTED** ESTIMATES MUSD, January 1990
CANADA	145 (HWR)	198
FINLAND	105 (PWR)	208
GERMANY	119 (PWR) 173 (BWR)	208 303
SWEDEN	107 (PWR) 140 (BWR)	198 259
UNITED STATES	97 (PWR) 113 (BWR)	121 141

*   Original estimates from countries had been modified to correspond to 1300 MWe unit size and to include contingency at 25 % in 1986 report.

**   (Adjusted estimate) = (Original estimate) x { (Consumer Price Index for 1990 January) x (exchange rate against US dollar for 1990 January) }/ { (Consumer Price Index for 1984 January) x (exchange rate against US dollar for 1984 January) }.

Table 18. Comparison of decommissioning cost estimates with waste arisings

ESTIMATE	REACTOR TYPE AND NUMBER	COST ESTIMATE (M$)	TOTAL PLANT CAPACITY (MW)	COST PER CAPACITY ($/W)	AMOUNT OF RADIOACTIVE WASTE* (Kton)	ADJUSTED COST ESTIMATE** (M$)	ADJUSTED COST PER UNIT OF WASTE (k$/ton)	COST PER PLANT UNIT (M$)
(LWRS)								
FI-1	PWR x 2	237	930	0.25	7.7	237.0	30.8	119
FI-2	BWR x 2	198	1 470	0.13	11.2	198.0	17.7	99
GER-1	PWR x 1	218	1 204	0.18	10.3	174.4	16.9	218
GER-2	PWR x 1	205	1 204	0.17	10.3	164.0	15.9	205
J-1	PWR x 1	225	1 160	0.19	11.8	216.0	18.3	225
J-2	BWR x 1	234	1 100	0.21	11.3	224.6	19.9	234
SW-1	BWR x 1	152	780	0.19	5.5	121.6	22.1	152
SW-2	PWR x 1	130	920	0.14	4.1	104.0	25.4	130
UK-4	PWR x 1	418	1 155	0.36	12.5	334.4	26.8	418
US-1	PWR x 1	120	1 175	0.10	8.5 **	120.0	14.1	120
(GAS-COOLED REACTORS)								
SPA	GCR x 1	410	500	0.82	14.3	393.6	27.5	410
UK-1	GCR x 8	1 488	480	3.10	81.4	1 071.4	13.2	186
UK-2	GCR x 2	1 048	438	2.39	37.8	754.6	20.0	524
UK-3	AGR x 2	993	1 320	0.75	23.8	715.0	30.0	497
(HEAVY WATER REACTORS)								
CAN	CANDUx1	238	600	0.40	8.3 ***	190.4	22.9	238
(ALL REACTORS) MAX/MIN				31			2.3	5.3
(LWRS ONLY, EXCLUDING FI-1) MAX/MIN				3.6			1.9	4.2

*   Radioactive waste arising from whole decommissioning project (not waste amount per plant).

**   Adjusted as follows:
  - The cost estimates which include the costs for dismantling non-radioactive parts are multiplied by 0.8.
  - The cost estimates which exclude the costs for waste handling and disposal are multiplied by 1.2.
  - UK-1, UK-2 and UK-3 are multiplied by 0.9 because these cost estimates include significant costs for defuelling.

*** Estimated by the Secretariat.

Figure 1. SCHEMATIC DESCRIPTION OF THE THREE DECOMMISSIONING STAGES FOR NUCLEAR POWER PLANTS
(including the demolition of buildings)

Stage 1

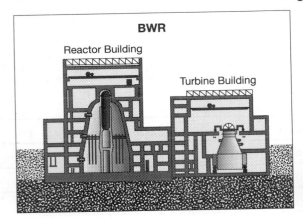

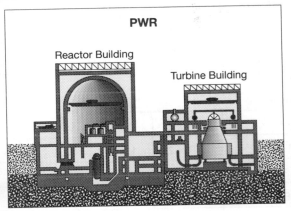

Stage 2

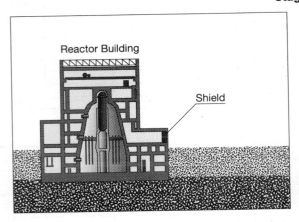

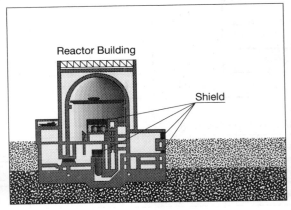

Stage 3

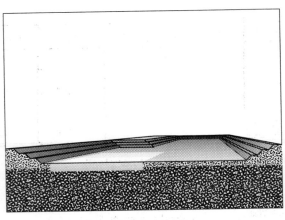

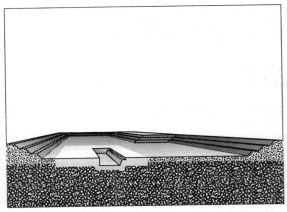

Note: The current Japanese decommissioning strategy assumes a short (5-10 years) storage period in Stage 1 before final dismantling of the plant to Stage 3.

Source: Ministry of Internationl Trade and Industry, Japan.

Figure 2. **EXCHANGE RATE * TRENDS**
For selected OECD Country currencies

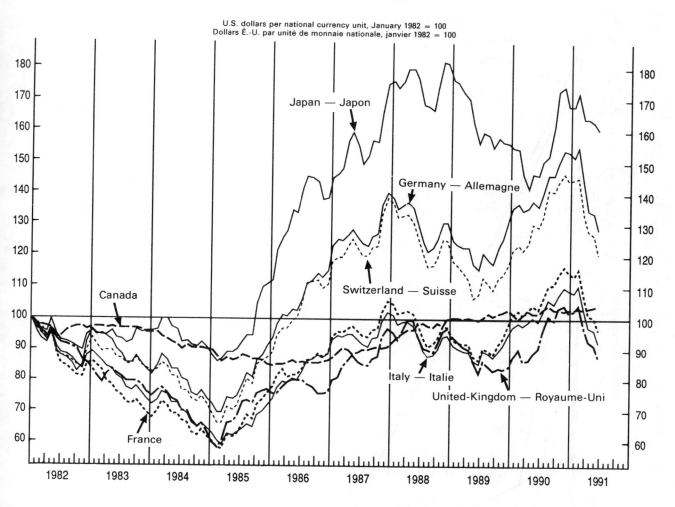

U.S. dollars per national currency unit, January 1982 = 100
Dollars É.-U. par unité de monnaie nationale, janvier 1982 = 100

Japan — Japon

Germany — Allemagne

Canada

Switzerland — Suisse

Italy — Italie

France

United-Kingdom — Royaume-Uni

1982 1983 1984 1985 1986 1987 1988 1989 1990 1991

* Daily average of spot rates quoted for the US dollar on national markets.

Origin: Main Economic Indicators (April 1991, OECD), Paris.

Figure 3. **RADIOACTIVE DECAY OF DEPOSITED INTERNAL SURFACE CONTAMINATION IN BWR**

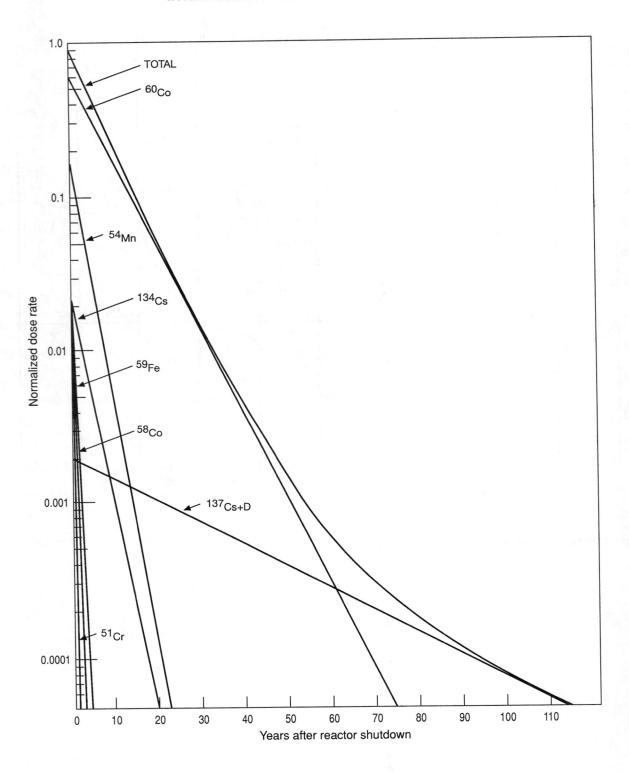

Origin: NUREG/CR-0672 Vol. 1 (1980)

Figure 4. MAJOR ISOTOPES WHICH CONTRIBUTE TO RADIATION EXPOSURE INSIDE A DECOMMISSIONED MAGNOX REACTOR

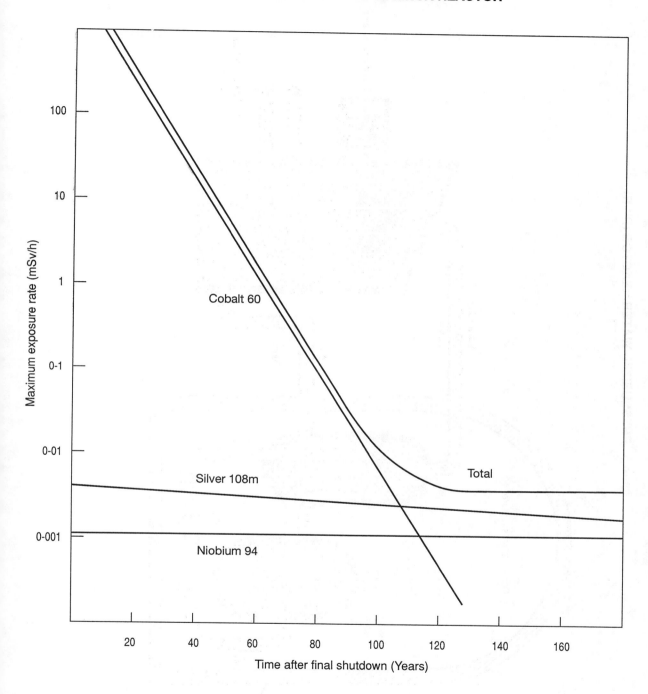

Only cobalt-60, niobium-94 and silver-108 m will contribute to radiation exposure inside a decommissioned Magnox reactor. After about 90 years only niobium-94 and silver-108 m will be significant.

Origin : P.B. Woollam, "Decommissioning Nuclear Power Stations", p. 106-113, The Nuclear Engineer, Vol. 30, No. 4.

Figure 5. **PRESSURIZED WATER REACTOR (PWR)**

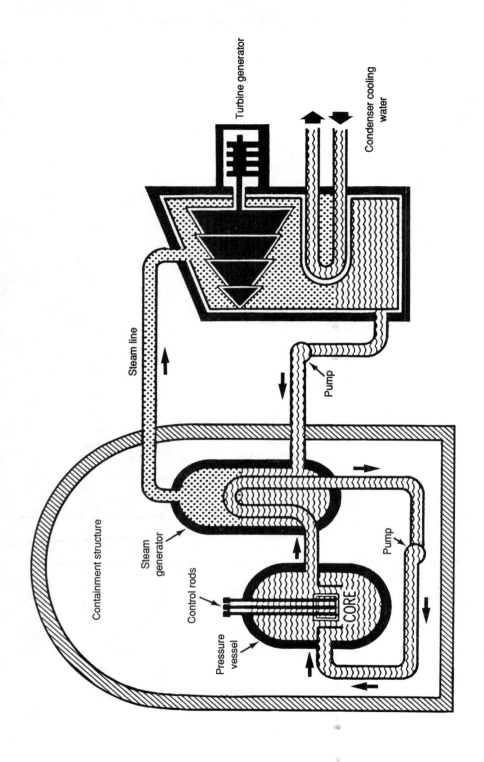

Turbine generator

Condenser cooling water

Steam line

Pump

Containment structure

Steam generator

Control rods

Pump

Pressure vessel

CORE

Figure 6. **BOILING WATER REACTOR (BWR)**

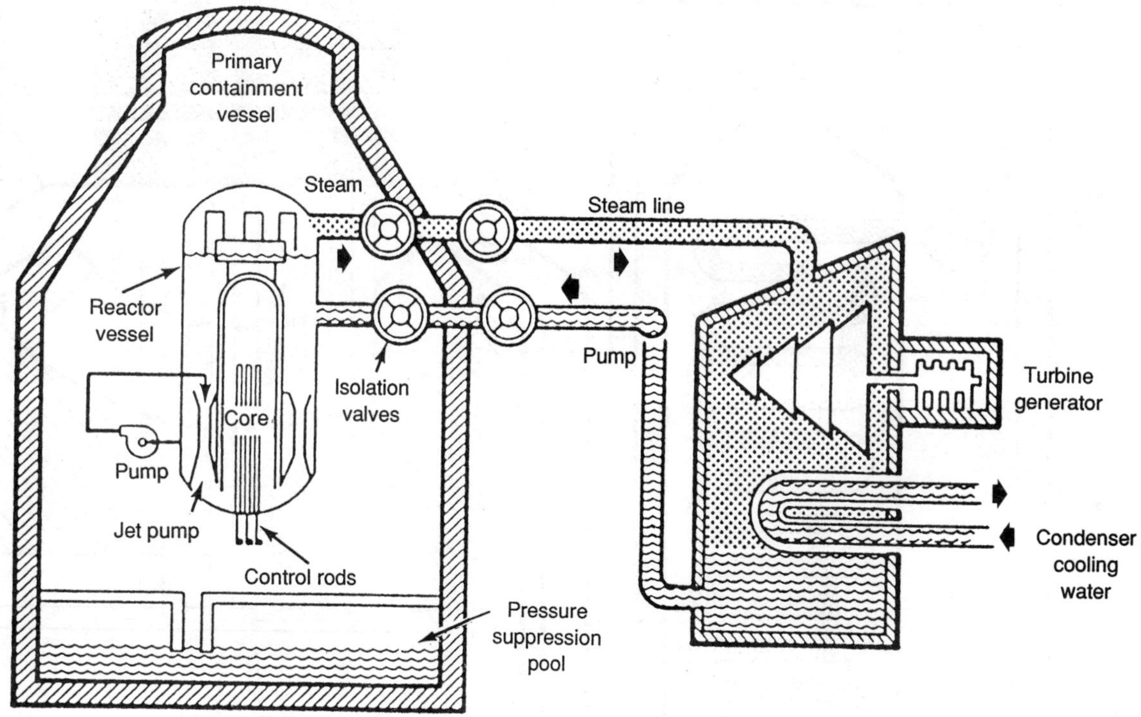

Turbine generator

Condenser cooling water

Steam line

Pump

Steam

Isolation valves

Pressure suppression pool

Primary containment vessel

Reactor vessel

Core

Control rods

Pump

Jet pump

Figure 7. **MAGNOX REACTOR**

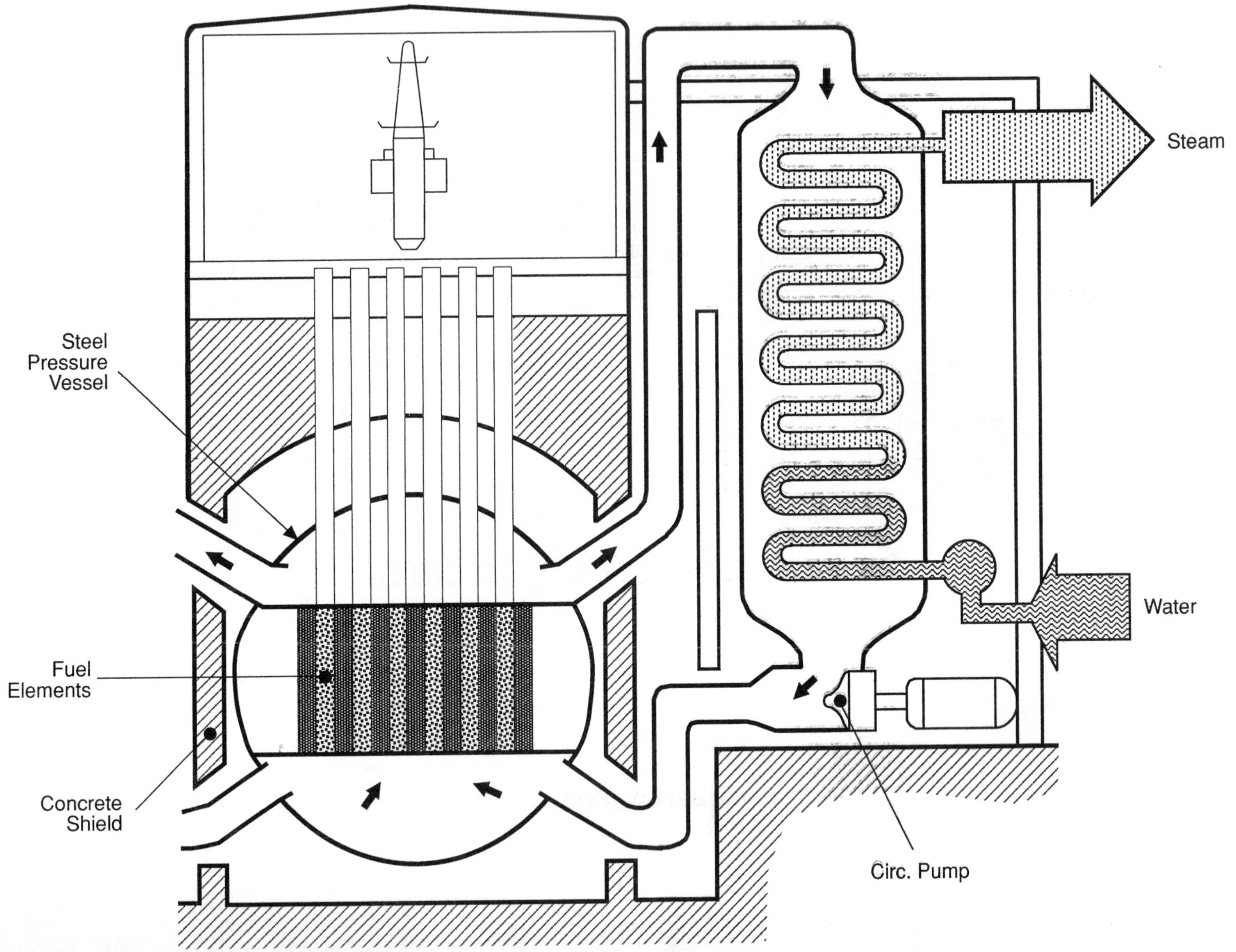

Steam

Water

Circ. Pump

Steel
Pressure
Vessel

Fuel
Elements

Concrete
Shield

Figure 8. **ADVANCED GAS COOLED REACTOR (AGR)**

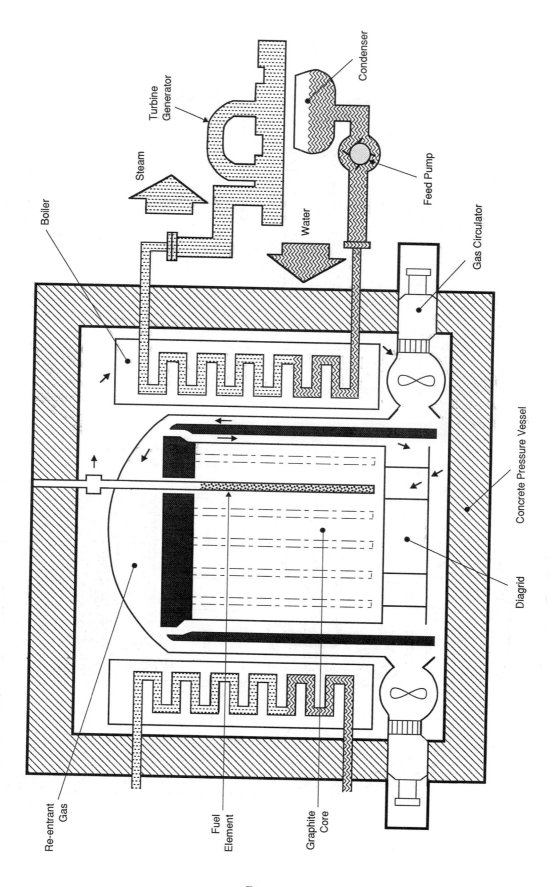

Turbine Generator

Condenser

Steam

Boiler

Water

Feed Pump

Gas Circulator

Concrete Pressure Vessel

Diagrid

Re-entrant Gas

Fuel Element

Graphite Core

Figure 9. **CANDU REACTOR**

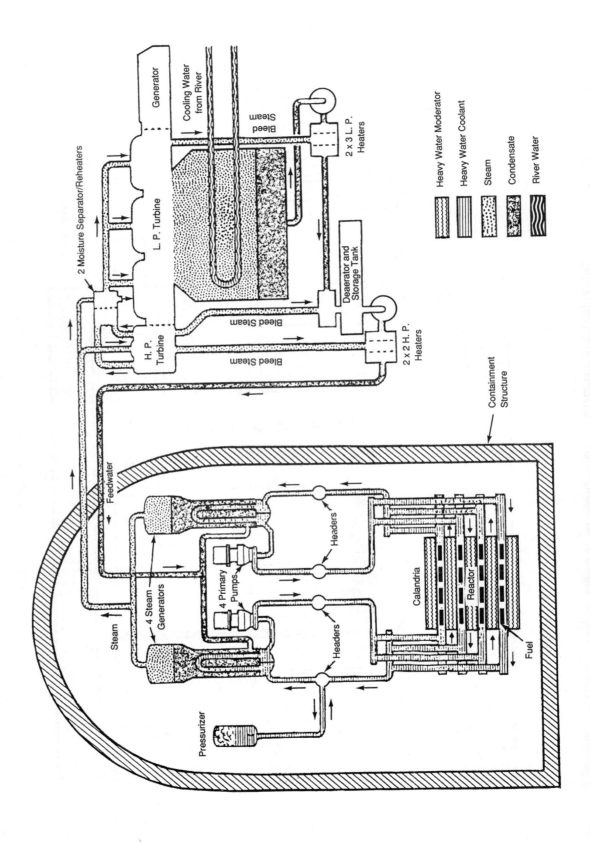

Figure 10. **TYPICAL DESIGNS OF BWR CONTAINMENT**

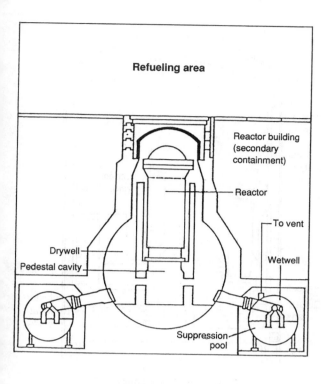

BWR MARK 1 CONTAINMENT

BWR MARK 2 CONTAINMENT

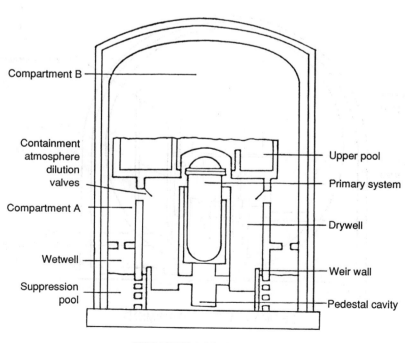

BWR MARK 3 CONTAINMENT

Figure 11. TYPICAL DESIGNS OF PWR CONTAINMENT

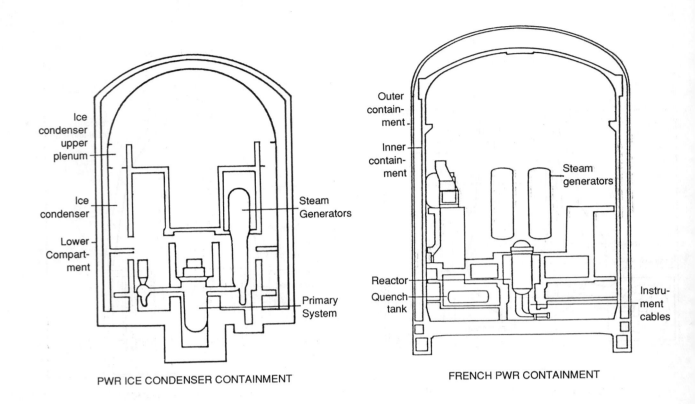

PWR ICE CONDENSER CONTAINMENT

FRENCH PWR CONTAINMENT

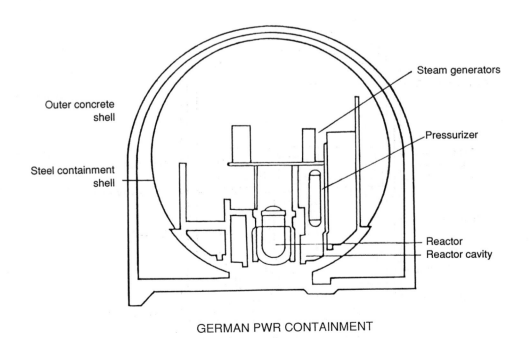

GERMAN PWR CONTAINMENT

74

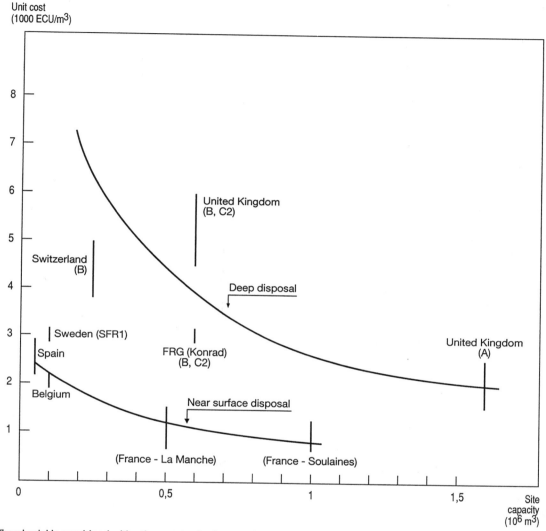

Figure 12. DISPOSAL OF CATEGORY "A" WASTE
UNIT COST EVOLUTION IN ACCORDANCE WITH SITE CAPACITY

1. When burial is combined with other waste, the latter are mentioned under the name of the country.

2. Waste categories are defined as follows:

   Category A: Conditioned low- and medium-level waste containing isotopes with a half-life not exceeding 30 years and possibly insignificant quantities of isotopes with a half-life exceeding 30 years;

   Category A: Conditioned low- and medium-level waste containing isotopes with a half-life exceeding 30 years;

   Category C: Conditioned high-level waste containing beta-gamma and alpha isotopes.
   As vitrified waste needs to be stored and disposed of in a specific way, it is treated separately. Therefore, the following distinction was considered necessary:
   C1: vitrified waste;
   C2: non vitrified waste.

Origin: Evaluation of storage and disposal costs for conditioned radioactive waste in several European countries, EUR 12871, CEC (1990).

DECOMMISSIONING COSTS FOR FUEL CYCLE FACILITIES

In response to the questionnaire which was circulated to the Expert Group members, several cost estimates were obtained for decommissioning fuel cycle facilities. A summary of these estimates is described in this annex.

The cost information from Belgium is for the decommissioning of the Eurochemic reprocessing plant. In addition to a cost estimate and its background, an explanation about the differences between the estimated costs and the real costs for a pilot decommissioning project is provided.

The CEA (Commissariat a l'Energie Atomique) of France provided the cost information for decommissioning a pilot reprocessing plant (AT-1). Some background information on the cost estimate is provided as well.

The cost information from the BNFL (British Nuclear Fuels plc.) includes five estimates (2 for storage/decanning facilities, 2 for parts of reprocessing plants and 1 for plutonium finishing lines). The cost information includes only the resultant cost estimates and brief breakdowns.

Table 1.1 is a summary table of the cost information included in this annex. The presentation follows the style in the Table 1 in the main text.

1.   Eurochemic reprocessing plant

Description of decommissioning project

The Eurochemic reprocessing facility, at the Mol-Dessel site in Belgium, was constructed in the early sixties (1960-1966). It was owned by a 13-nation consortium and went in active operation in 1966 as a demonstration plant. The plant had a throughput of 60 tons per year. Between 1966 and 1974, approximately 180 tons of low enriched uranium fuel and 30.5 tons of highly enriched uranium fuels were reprocessed. The plant was shutdown in January 1975.

During the years 1975-79, the reprocessing plant was decontaminated for keeping it in a safe standby condition at a reasonable cost. The plant entered a dormancy period from 1980. The plant has been transferred in stages to Belgium and has fully been owned by Belgium since 1983. Belgoprocess was established to take charge of the activities on site at the end of 1986 and is a subsidiary of NIRAS/ONDRAF.

As from 1987, the Eurochemic site has been used as the centralized interim storage for all conditioned radioactive wastes, and is also gradually being used for centralized waste processing. For this the site will have to be re-modelled and many of the buildings that were erected for specific purposes have to be dismantled. Among the buildings identified for dismantling during the first phase are:

- the reprocessing plant itself,
- its analytical laboratory,
- the storage facility for the end products of reprocessing,
- storage tanks for high and intermediate levels liquid wastes.

The decommissioning activities started in 1987 with a pilot project. Decommissioning of the reprocessing plant itself started in 1989 and it can be expected that it can be executed in about 15 years.

Cost estimate and its backgrounds

A preliminary estimate, which was performed in 1987, indicated that the dismantling of the above facilities would cost about BF 5 725 million (1987 value). This is an undiscounted cost and includes a contingency allowance (equivalent to 30 per cent of the "best estimate").

As shown in Table 1.2, the estimate covers all costs starting from the preparation of decommissioning plan to site clean-up and landscaping. Not included were the cost for fuel back-end and the cost for conventional demolition of the main building.

For the estimate, all equipment, material and areas with contamination levels above 4 Bq/cm^2 beta-gamma and 0.4 Bq/cm^2 alpha were considered radioactive. However, later on during the first decommissioning works for the pilot project, the exemption levels were lowered to 0.4 Bq/cm^2 beta-gamma and 0.04 Bq/cm^2 alpha. As a result, all surface area had to be monitored at 100 per cent. Surfaces/areas that could not be monitored were considered radioactive. Additionally, core samples of concrete structures had to be taken on the previously most contaminated spots. The specific activities of these samples had to be well below 1 Bq/g. Material monitored and found to be under the criteria could be disposed of without any restrictions. Table 1.3 shows the radioactive waste classification system adopted in the cost estimate.

Radioactive waste will be treated and disposed of to the central intermediate storage facility located on the same Eurochemic site. Final disposal of the conditioned waste will be by shallow land burial (for low radioactive wastes with tentatively accepted limit value for alpha content lower than 3.7 GBq/ton conditioned waste) and by deep geological disposal (for the other radioactive wastes with alpha content higher than the tentatively accepted limit value of 3.7 GBq/ton conditioned waste). Unit costs for waste disposal assumed for the estimate were:

- shallow land burial:
    - intermediate disposal   BF   30 000 per cubic meter
    - final disposal   BF   85 000 per cubic meter
- geological disposal:
    - intermediate disposal   BF   75 000 per cubic meter
    - final disposal   BF   600 000 per cubic meter

Table 1.4 shows the quantities of various radioactive wastes based on the original exemption levels. In the table, primary wastes are those removed from the facilities dismantled, and secondary wastes are those produced in the decommissioning activities (e.g. decontamination liquid, working clothes, ventilation filters, etc.).

It was estimated that the total labour requirements would be 835 man-years. The total manpower cost represents approximately 54 per cent of total decommissioning costs estimated. The allowable occupational dose limit to the workers assumed in the estimate was a maximum of 30 mSv/year instead of the legal limit of 50 mSv/year (more recently the legal limit has been lowered to 20 mSv/year).

Comparison of estimated costs and real costs for pilot decommissioning project

Two buildings, which had been used for the storage of uranyl nitrate and spent solvents at the Eurochemic site, have been emptied and decontaminated as a pilot project to check techniques and costs, as well as to train personnel before embarking on the dismantling of the major part of the plant. In order to demonstrate the feasibility of the general decommissioning strategy, the demolition of the building and the removal from the site are included as project objectives.

Table 1.5 is a comparison table of the estimated costs and the real costs for the pilot decommissioning project. Although the real costs for the item "Dismantling Activities" are higher than the estimated costs, the total real costs shows a good agreement with the total estimated costs. Owing to adaptation in the decontamination work, the real quantity of resulting waste was limited to only 40 per cent of earlier estimates.

One of the reasons for the higher costs for "dismantling activities" can be found in the character of the pilot project itself. It required additional costs to check the decommissioning technology and collect cost data, and to train the inexperienced operators. Another reason was the fact that the free release level was lowered, during the decommissioning period, from 4 Bq/cm² beta-gamma to 0.4 Bq/cm² beta-gamma and from 0.4 Bq/cm² alpha to 0.04 Bq/cm² alpha. These changes led to more decontamination and more scabbling work as well as 50 per cent more monitoring for release (costs for about BF 3.6 million higher). The higher costs for this work were partially compensated by the lower resulting quantity of waste production and hence the lower waste management costs.

2.   AT-1

AT-1, located at the La Hague plant, France, was a pilot plant for reprocessing of mixed oxide fuel elements from fast breeder reactors (mainly from Rapsodie and partly from Phenix). The plant construction started in 1964 and it operated successfully from 1969 to 1979. The plant had a throughput of 1 to 2 kg-HM/day, using the PUREX process.

AT-1 plant is being decommissioned to Stage-3 and its building will be released for unrestricted use. The main decommissioning steps are the following:

•   Dismantling of unshielded cells and glove boxes of the end cycle of the PUREX process. These cells and glove boxes were contaminated with alpha emitters but not so highly with beta-gamma emitters. For this operation a reusable modular alpha-tight work-shop was built.

•   Dismantling of accessible hot cells. The hot cells (including the blind hot cells mentioned below) are contaminated not only with alpha emitters but also highly with beta-gamma emitters, and so are heavily shielded.

•   Dismantling of main shielded blind hot cells, which were not equipped with any shielded windows or manipulators. A remote controlled machine (ATENA), which has a 6 meter long articulated arm for dismantling the blind cells, was designed and was installed on the floor of the accessible hot cells which had been dismantled.

•   Dismantling of cells housing various storages (effluents, fission products...).

•   General cleaning of the building for unrestricted use.

The decommissioning project, which started in 1981, is on schedule and will be completed in 1992.

Cost estimate and its backgrounds

The decommissioning costs of this project, estimated in 1989, were 220 million Francs (1989 French Francs). This is an undiscounted cost and includes contingency allowance (equivalent to 4.6 per cent of "best estimate"). Table 1.6 is a summary of the estimate. The estimate includes the costs for preparation of decommissioning plan, the costs for site cleanup and landscaping and costs for research & development.

The building demolition after decontamination is not included in the scope of the project. The cost estimate excludes shut-down costs and fuel back-end costs.

Radioactive wastes removed will be 2 500 cubic meters after packaging for disposal. The criterion for free release was assumed as 1 million Bq per ton. The radioactive wastes were classified into three categories (see Table 1.7). Disposal of low level wastes was by shallow land disposal (surface trenches and monoliths). The disposal site has been the Site de Stockage de la Manche (SSM), which is located nearby the La Hague plant and will be closed in 1991. A new site, the Centre de Stockage de l'Aube (CSA), will be opened in 1991 near Troyes in the center of France. The intermediate level and high level wastes will be disposed of to a geological disposal site, which will be constructed in the future.

Unit costs for low level radioactive waste (LLW) treatment were assumed FF 2 800 per cubic meter; unit costs for LLW transportation were assumed FF 600 per cubic meter; unit costs for LLW disposal to a shallow land disposal site were assumed FF 5 000 per cubic meter.

The total labour requirements were assumed as 323 000 man-hours (Direct hands-on labour, 232 000 man-hours; Overhead and administration, 40 000 man-hours; Engineering, 51 000 man-hours). The total labour costs were considered to be FF 250-600 per hour (Wages/salaries, FF 150-450/h; other labour related costs, FF 100-150/h). The regulatory occupational dose limit to workers is 50 mSv/year, but a target limit in practice was assumed as 5-10 mSv/year.

3. BNFL's fuel cycle facilities

BNFL has undertaken an assessment of all its decommissioning liabilities covering plants already shut down, operating, under construction or only at the planning stage. This has led to a total liability assessment of £4.6 billion and a 100 year programme. Because of the number of plants to be assessed, a factoring system was used. For new plants the decommissioning costs were assumed to be a proportion of construction costs whilst for older plants they were factored from known costs for plant modifications, etc., combined with a difficulty element for the particular plant. The study yielded global costs but the next phase is to undertake more detailed assessments for particular plants. The plants fuel cycle facilities described below are included in this more detailed assessment. It should however be emphasized that none of them are due for imminent decommissioning indeed some have not even started operation. The details from the studies are therefore extremely limited compared with a detailed decommissioning plan for a plant due to be decommissioned shortly. For each plant there has been a more detailed assessment of the waste volumes and classification and the amount of dismantling required. For large plants this has involved detailed study of representative areas only, scaled up for the whole plant.

3.1 B-30 Pond

This plant operated for 30 years from the mid 1960s receiving, storing and decanning over 30 000 tonnes of Magnox spent fuel from the UK, Japan and Italy. The facility is split into a receipt building, the cooling ponds and the decanning building. The receipt building is 33m x 23m x 19m high with reinforced concrete walls 1m thick containing three fuel flask delidding and skip transfer caves, flask decontamination and associated mechanical handling and other services. The cooling and storage ponds are approximately 145m x 17m x 5.4m deep with 1m reinforced concrete walls. The ponds are uncovered and served by two skiphandler machines for movement of the fuel skips. There has been corrosion of some of the fuel over the years and there is a substantial sludge layer including windblown and organic debris. The decanning building is now some 145m x 15m x 20m high. It contains six underwater decanning bays together with underwater maintenance bays. The underwater bays were superseded by two dry decanning caves in an extension to the original building some 48m x 3m x 10m high with reinforced concrete walls 1m thick. The plant ceased its main operation role in the later 1980s and is currently undergoing post operational clean out (POCO) including decanning of residual fuel and the removal of accumulated sludges and debris, a programme which will continue for several years.

The decommissioning activities will follow the completion of POCO and are currently estimated at of £44 million (1988 value, undiscounted). Table 1.8 shows a breakdown.

3.2   B-311 fuel handling plant

This plant became operational early in 1986 and has largely replaced B-30 for Magnox fuel whilst also storing AGR fuel elements and breaking them down into pins to allow more efficient use of storage space. The Magnox fuel is decanned in the plant prior to shipment to the B-205 reprocessing plant whilst the broken down AGR fuel will be transferred to THORP for reprocessing.

Unlike B-30, the plant is completely covered and divided into receipt, storage and decanning/breakdown areas. The overall plant is 160m x 90m x 30m high. The receipt section contains three inlet cells each 40m x 10m x 9m high with walls 1.5m thick. The storage section has three main ponds each 30m x 30m x 8m deep and three sub ponds provide various skip wash and fuel preparation facilities. The fuel is stored in skips within containers unlike B-30 where it is stored only in skips. The decanning/breakdown section contains two decanning cells 64m x 8m x 10m high and 50m x 8m x 10m high and an AGR fuel breakdown cell some 64m x 10m x 15m high. The cell walls are 1.5m thick. The plant in addition has a flask maintenance cell, Calder fuel basket decontamination cell and various decontamination and other maintenance cell together will all the associated services.

The current estimate of undiscounted costs for decommissioning B-311 is £140 million (1988 value). Table 1.9 shows a breakdown.

3.3   B-205 chemical separation plant

This is the Magnox reprocessing plant and has been operating since the mid 1960s with a planned life of a further 20 years. The plant receives decanned Magnox fuel from B-311 which is then dissolved in nitric acid and separated into its various streams using the PUREX process. The uranium and plutonium streams are sent for delay storage or vitrification. The plant consists of two dissolver, various solvent and service cells all enclosed in a linked concrete structure with wall thickness up to 1m. The plant has processed over 2 500 tonnes of spent fuel and has undergone modification and refurbishment.

The current estimate of undiscounted costs for decommissioning B-205 is £52.5 million (1988 value). Table 1.10 shows a breakdown.

3.4   THORP head end and chemical separation plant

This is the main new reprocessing plant for Sellafield for reprocessing oxide fuels from AGR, PWR and BWR stations. The plant has two fuel storage ponds, fuel shearing and dissolution cells and the main chemical separation facility including plutonium and uranium finishing. The plant is massive approximately twice the size of B-311 with a design throughput of 1 200 tonnes per year. The plant has integral decontamination, maintenance and ventilation facilities. Decommissioning was considered during the design and construction phase of this plant.

The current estimated decommissioning costs are £403.6 million (1988 value, undiscounted). Table 1.11 shows a breakdown.

3.5 Plutonium finishing line 4 and 5

Finishing line 4 and 5 have been operated since the early 1980s and convert plutonium nitrate to oxide. Details of their layout and throughput are confidential though they are typical of glovebox plutonium facilities. The total undiscounted cost of their decommissioning is estimated to be £27.0 million (1988 value). Table 1.12 shows a breakdown.

Table 1.1 Decommissioning cost estimates for fuel cycle facilities

FACILITY NAME	FACILITY DESCRIPTION			MODES OF DECOMMIS-SIONING	ESTIMATED DECOMMISSIONING COSTS		YEAR OF ESTIMATE	CPI(b) ESCALATOR APPLIED	EXCHANGE RATE OF JAN. 1990 (NCUs per USD)
	Start of construction	Start of operation	End of operation		As given in the response to the questionnaire (a)	in millions of US $ of Jan. 1990			
Eurochemic reprocessing plant (BELGIUM)	1960	1966	1974	10 years + Stage-3	5 724.6 M BF(1987)	172.9	1987	1.071	35.47
AT-1 pilot reprocessing plant (FRANCE)	1964	1969	1979	Stage-3	220 M FF(1989)	39.5	1989	1.034	5.761
B30 storage/decanning pond (UK)	-	mid 1960s	-	-	44.4 M £ (1988)	85.0	1988	1.158	0.605
B311 fuel handling plant (UK)	-	1986	-	-	140.1 M £ (1988)	268.2	1988	1.158	0.605
B205 Chemical Separation plant (UK)	-	mid 1960s	-	-	52.5 M £ (1988)	100.5	1988	1.158	0.605
THORP head end and chemical separation plant (UK)	-	1992	-	-	403.6 M £ (1988)	772.5	1988	1.158	0.605
Plutonium finishing lines	-	early 1980s	-	-	27.0 M £ (1988)	51.7	1988	1.158	0.605

(a) The number in parenthesis is the base year for the money value.
(b) Calculated as x/y   x = the national consumer price index (CPI) for January 1990
        y = the national CPI for January of the base year of money value.

83

Table 1.2 Decommissioning cost estimates for Eurochemic reprocessing facility

(in 1987 Million Belgian Francs)

ITEMS	COST
1. Preparation of Decommissioning Plan	37.5
2. Procurement of Equipment and Material	122.9
3. Security, Surveillance and Maintenance	1 020.7
4. Preliminary Decontamination	180.0
5. Dismantling Activities	1 723.4
6. Waste Management and Quality Control	900.3
7. Site Clean-up and Landscaping (exclusive main building, set to free release and re-use as a storage building)	29.2
8. Intermediate Waste Disposal	96.3
9. Final Waste Disposal	293.2
Sub-total	4 403.5
Contingency (30 %)	1 321.1
TOTAL	5 724.6

Table 1.3 **Radioactive waste classification system adopted in Eurochemic decommissioning cost estimate**

WASTE CATEGORY	CRITERIA			
A. Liquid waste				
- low radioactive liquid waste:	max. β-γ		max. α	
- suspected condensates	0.04	MBq/m^3	0.004	MBq/m^3
- cold liquid waste	0.4	MBq/m^3	0.04	MBq/m^3
- low liquid waste	400	MBq/m^3	0.8	MBq/m^3
- warm liquid waste	40 000	MBq/m^3	80	MBq/m^3
- middle radioactive liquid waste	20 000	GBq/m^3	8 000	GBq/m^3
- high radioactive liquid waste	6 500	TBq/m^3	30	TBq/m^3
B. Non-liquid wastes				
- low radioactive, non-liquid, beta-gamma waste:	* beta-gamma activity : < 40 GBq/m^3 * no Pu, Ra or Th * U : < 20 MBq/m^3 * other alpha activity : ≤ 40 MBq/m^3 * dose rate : < 2 mSv/hr each-package			
- low radioactive, non-liquid, alpha-suspected, beta waste:	* beta-gamma activity : < 40 GBq/m^3 * no Ra, Th * U : < 20 MBq/m^3 * alpha activity : < 4 GBq/m^3 * dose rate : < 2 mSv/hr each-package			
- low active, non-liquid alpha-waste:	* beta-gamma activity : ≤ 40 GBq/m^3 * alpha activity : > 4 GBq/m^3 * dose rate : < 2 mSv/hr each-package			
- medium level non-liquid waste:	* dose rate : < 2.5 Sv/hr			
- high level non-liquid waste:	* dose rate : < 16 000 Sv/hr			

**Table 1.4 Quantities of radioactive wastes estimated for Eurochemic reprocessing facility
(based on the original exemption level*)**

A. PRIMARY WASTES

  - for deep geological disposal
    (medium and high level wastes, tentatively accepted limit for alpha content higher than
    3.7 GBq/t conditioned waste)

> • metal waste   : 230.6 tonnes
> • concrete waste : 30   tonnes
> • special waste   : 12 m^3

  - for shallow land disposal
    (low level wastes, tentatively accepted limit for alpha content lower than 3.7 GBq/t conditioned
    waste)

> • metal waste   : 1 095.3 tonnes
> • concrete waste : 3 235   tonnes
> • special waste   :   180 m^3

B. SECONDARY WASTES
   (low level wastes)

> • cold liquid waste   : 40 083 m^3
> • low level liquid waste : 14 171 m^3
> • burnable waste   :   903 m^3
> • compactable waste   : 10 412 filters

*   4 Bq/cm² beta-gamma and 0.4 Bq/cm² alpha.

Table 1.5 Comparison table of estimated costs and real costs for pilot decommissioning project

(in 1987 Million Belgian Francs)

ITEMS	ESTIMATED COSTS (M BF)	REAL COSTS (M BF)
1. Preparation and Decommissioning Plan	0.75	3.460
2. Procurement of Equipment and Material	2.46	6.844
3. Maintenance of Systems in Operation	5.60	0.785
4. Preliminary Decontamination	-	0.466
5. Dismantling Activities	33.90	50.765
6. Waste Management and Quality Control	18.00	8.501
7. Site Clean-up	1.46	2.863
8. Intermediate Waste Disposal	1.93	1.266
9. Final Waste Disposal	5.86	6.326 *
10. Project Management	2.92	1.837
Sub-total	72.88	83.113
Contingency (30 %)	21.86	-
TOTAL	94.74	83.113

* Real figures included costs for disposal of demolition waste of released concrete structure on an industrial dumping ground.

**Table 1.6 Decommissioning cost estimate for AT-1 reprocessing plant
(in 1989 Million French Francs)**

ITEMS	COST (M FF)	%
1. Preparation and Decommissioning Plan	2	0.9
2. Procurement of Equipment and Material	40.5	18.4
3. Preliminary Decontamination and Dismantling Activities	28	12.7
4. Waste Management and Quality Control	8.5	9.3
5. Disposal Cost	12	
6. Security, Surveillance and Maintenance	73	33.2
7. Site Clean-up and Landscaping	3	1.4
8. Project Management, Engineering and Site Support	35	15.9
9. Research and Development	2	0.9
10. Overheads	6	2.7
11. Contingency	10	4.6
TOTAL	220	100

**Table 1.7 Radioactive waste classification system adopted in
AT-1 decommissioning cost estimate**

WASTE CATEGORY	CRITERIA

Category A (low level waste)	- Imbedded waste
	   < 3.7 GBq/t alpha
	  $f^* < 10$ beta-gamma
	
	- Not-imbedded waste
	   < 0.2 GBq/t alpha
	   < 37 GBq/t beta-gamma

Category B (intermediate level waste)	> 3.7 GBq/t alpha
Category C (high level waste)	Vitrified waste

* $f =$ Summation of (ai/LMAi) for all beta-gamma emitters
    ai   = activity of nuclide "i"
    LMAi = limit of mass activity for nuclide "i"

Table 1.8 Decommissioning cost estimate for B-30 pond
(in 1988 Million Pounds)

ITEMS	MANPOWER (M£)	WASTES (M£)	TOTAL (M£)
Initial Decommissioning	9.4	6.1	15.5
Plant Dismantling	16.5	2.6	19.1
Building Demolition	6.3	3.0	9.3
Care and Maintenance	0.5	0	0.5
TOTAL	32.7	11.7	44.4

Table 1.9 Decommissioning cost estimate for B-311 fuel handling plant
(in 1988 Million Pounds)

ITEMS	MANPOWER (M£)	WASTES (M£)	TOTAL (M£)
Initial Decommissioning	19.9	1.4	21.3
Plant Dismantling	61.1	13.0	74.1
Building Demolition	39.1	0	39.1
Care and Maintenance	5.6	0	5.6
TOTAL	125.7	14.4	140.1

Table 1.10 Decommissioning cost estimate for B-205 chemical separation plant
(in 1988 Million Pounds)

ITEMS	MANPOWER (M£)	WASTES (M£)	TOTAL (M£)
Initial Decommissioning	3.4	1.1	4.5
Plant Dismantling	21.7	10.5	32.2
Building Demolition	10.0	5.7	15.7
Care and Maintenance	0.1	0	0.1
TOTAL	35.2	17.3	52.5

Table 1.11 Decommissioning cost estimate for THORP head end and chemical separation plant (in 1988 Million Pounds)

ITEMS	MANPOWER (M£)	WASTES (M£)	TOTAL (M£)
Initial Decommissioning	37.4	5.6	43.0
Plant Dismantling	183.0	50.6	233.6
Building Demolition	66.5	0	66.5
Care and Maintenance	60.5	0	60.5
TOTAL	347.4	56.2	403.6

Table 1.12 Decommissioning cost estimate for plutonium finishing lines No. 4 and 5 (in 1988 Million Pounds)

ITEMS	MANPOWER (M£)	WASTES (M£)	TOTAL (M£)
Initial Decommissioning	5.9	0.4	6.3
Plant Dismantling	14.4	3.4	17.8
Building Demolition	2.2	0	2.2
Care and Maintenance	0.7	0	0.7
TOTAL	23.2	3.8	27.0

Annex 2

COSTS FOR ONGOING/COMPLETED DECOMMISSIONING PROJECTS

Many decommissioning projects have been started in Member countries and some of them have been completed. Their main purposes were the research and development of decommissioning technology, and the demonstration of feasibility of decommissioning. The technical experience obtained in these projects is fedback to future decommissioning planning. The experience has proved the feasibility of decommissioning techniques and reduced the uncertainty in cost estimates.

Certainly, the costs of these ongoing or completed projects are not directly comparable with the future decommissioning costs of commercial power plants as explained in Chapter 4. However, in addition to various cost data obtained from these projects, experience of cost estimation for these projects is very valuable for the cost estimators of future decommissioning.

This annex provides cost information of recent projects which were provided to this study as responses to a questionnaire prepared by the Secretariat. Included are:

- Shippingport, USA, 72 MWe PWR
- Gentilly-1, Canada, 250 MWe CANDU
- KKN-1, Germany, 100 MWe HWGCR
- Chinon-A2, France, 250 MWe GCR

Table 2.1 is a summary of the cost information of these decommissioning projects. The table follows the presentation structure in the Table 1 of the main text. The outlines and some background cost information of these projects are described in following section.

1.   Shippingport decommissioning project

Description of the project

The Shippingport Atomic Power Station, which was located on Ohio river in Pennsylvania, was constructed from 1955 under the Eisenhower Atoms For Peace Programme. The station began its commercial operation from December 1957 and was operated by a public utility, Duquesne Light Company, under supervision of the U.S. Department of Energy-Naval reactors until operations ceased in October 1982. The station was a Westinghouse PWR and its design gross capacity was 72 megawatt electric average. Over the 25 year operating life of the station, there were 80 323 effective full power hours.

The decommissioning project was organised as a demonstration of the safe and cost effective dismantling of commercial scale nuclear power plant and to return the site to the owner for use with no radiological restriction. The station was first brought to Stage-1 and stored for 3 years, then finally dismantled to Stage-3 (September 1985 to October 1989). All physical decommissioning work on site was completed on July 1989, about 6 months ahead of schedule. The approval for release of site was issued in December 1989.

The DOE estimated that decommissioning the plant would cost $98.3 million (around 0.1 cent/kWh). In fact, the final cost was 10 per cent less, i.e. $91.3 million. The actual experienced occupational dose was 155 man-rem (1.55 manSv) to be compared with an estimated specific occupational dose of 1 007 man-rem (10.07 manSv).

Backgrounds of decommissioning cost of Shippingport

This section provides the actual cost information which were compiled in the Shippingport project.

Labour requirements

There were approximately 1 350 000 man-hours expended on project activities including planning activities prior to commencement of operations. The total project work force staff reached an average of around 200 during the peak activity years of 1986 and 1987. The head counts taken in September of each year for manpower levels for only physical decommissioning direct hands-on work are: 1984, 46; 1985, 163; 1986, 183; 1987, 238; 1988, 119; 1989, 2; and 1990, 0. The direct hands-on labour force man-hours was approximately 76 per cent of the total. The decommissioning operations and project management (including technical management support) man-hours was approximately 22 per cent. The overhead and administration man-hour is considered to be approximately 2 per cent of the total.

Labour rates

The hourly rate for a project laborer can be considered as a rate of $12.75 per hour plus 23.6 per cent ($3.01 per hour) for benefits and normal taxes. In addition, a surcharge of 20 per cent was assessed for radiation workers' workman's compensation insurance.

The labour hourly rate was subject to a legally mandated process that in 1985 resulted in a rate of $16.62 and which in 1987 resulted in a rate of $15.75. The lower rate's impact was limited because by 1987 many contracts had been awarded at the higher rate. The higher labour rate applied after project start-up is judged to have increased project costs by a minimum of 2 per cent.

An allowance was made for tools that became radioactively contaminated and disposed of with other contaminated waste. The percent was 7.4 % and included as a factor in functional work estimates.

Amounts and classification of radioactive waste

The physical decommissioning consisted of the demolition and disposal of various fluid and electrical systems before the buildings could be demolished. The project processed approximately 2 287 600 l (602 000 gallons) of radioactive liquid waste. Approximately 245 m^3 (320 cubic yards) of contaminated in site asbestos insulation was removed. Concrete scabbling removed contaminated in-situ concrete amounting to approximately 11 m^3 (15 cubic yards).

The reactor pressure vessel/neutron shield tank as removed was approximately 285 m^3 (375 cubic yards) and weighed approximately 870 tonnes (970 short tons). The package as prepared for shipment was approximately 275 m^3 (363 cubic yards) and weighed approximately 835 tonnes (930 short tons). During the project the total radioactive waste packaged for disposal was approximately 6 020 m^3 (7 900 cubic yards). Table 2.2 summarizes shipment data.

Concerning radioactive waste classification, U.S. Federal law requires radioactive waste transferred to a land disposal facility be classified in one of three classifications, Class A, Class B, and greater than Class C, according to concentrations of a specific group of nuclides; and further, for burial purposes all forms of radioactive waste shall adhere to the prescribed physical characteristics. Also, for transportation purposes, radioactive waste must be classified according to radiological and physical properties. Certain classifications for transportation are made by the quantity or concentration of radioactive material by specific radionuclide.

Criterion for release of the Shippingport site without radiological restriction was defined as 100 mrem (1 mSv) per year total committed effective dose equivalent to the maximum exposed individual of the general public under the worst case scenario plus the application of the philosophy of ALARA (as low as reasonably achievable) to 10 mrem (0.1 mSv) per year. The final predicted doses for the three key scenarios were: 2 mrem (0.02 mSv) per year for residual; 4 mrem (0.04 mSv) per year for buried concrete pipe chases; and 24 mrem (0.24 mSv) for deeply buried, but potentially occupiable areas.

Unit costs for treatment, transportation and disposal of radioactive waste

These unit costs were obtained by dividing the total costs for these activities by the amount of radioactive waste after packaging. The approximate total costs for liquid and solid wastes management amounted to $3.426 million (US 1989 dollars). A gross cost for radioactive waste treatment was $565.07 per cubic meter ($432.30 per cubic foot) including collection, processing, disposal packaging and disposal shipment.

Truck and rail shipments of radioactive waste were a distance of 3 800 kilometers (2 400 miles) and barge shipment was a distance of 13 400 kilometers (7 400 nautical miles). The total truck and rail waste transportation costs was $1.179 million (US 1989 dollars), and a gross cost per cubic meter was $203.98 (US 1989 dollars).

Disposal of Shippingport waste was by shallow land burial. All radioactive materials above acceptable residual levels were removed and transported to an existing radioactive waste disposal site at DOE Hanford Reservation. The Reservation assessed a burial cost of $5.92 per cubic foot in 1986 for low level waste. By 1989, the unit cost had increased to $27.06 per cubic foot. The total solid waste disposal cost was $2 438 000 over the life of the project for the 6 063 m^3 of waste. These data render an average gross disposal cost per cubic meter of $402.11 (US 1989 dollars) or an average gross disposal cost per cubic yard of $307.63 (US 1989 dollars).

Others

A major feature of the project was the one-piece removal of the reactor pressure vessel (RPV) and its shipment by barge to the Hanford Reservation in the State of Washington. Total cost to prepare to remove, and bury the package was $10.3 million (US 1989 dollars). Table 2.3 provides its allocations. In the table, "Task %" represents the proportion of the item cost to the total cost of the task (RPV lift, shipment and transportation) and "Project %" represents the proportion to the total cost for the decommissioning project.

When the decommissioning cost was estimated, a contingency allowance was included. The range was from 8 per cent to 20 per cent with an average of a 16 per cent contingency.

Explanation of differences between the estimates and actual experienced costs

Table 2.4 is a summary of the estimated and actual costs of Shippingport project. Only major project cost elements are summarized by project Work Breakdown Structure (WBS) element. Accumulated project costs, displayed in the final two lines of the table, are not intended to be obtained by direct addition of each preceding line entry in the columns. The last entry in the Project Cost Baseline column is the Total Estimated Cost developed from project planning. The Target Cost Baseline column is the negotiated contract modifications. The final Cost column is actual cost accrued through the end of the project.

Differences between column one and two are changes in scope of work. Differences between column two and three are differences in cost for equivalent scope of work. The data are US dollars gathered from project records accrued through 1989.

Table 2.5 displays the Work Breakdown Structure (WBS) elements with the highest percent of change over the Project Plan Cost Baseline.

a) Hanford burial service

The project Final Cost of approximately $2.5 million exceeded the original Project Plan Cost Baseline of $0.6 million. This change arose from extraordinarily large increase in the waste disposal costs. These increases were in burial cost per unit volume assessed at the DOE Hanford waste burial site. In 1986, the cost for low-level waste burial was $5.92 per cubic foot and had increased to $27.06 by 1989.

b) Liquid waste management

Liquid waste disposal operations increase was due mainly to the increase in the liquid waste volumes processed for adherence to the project management guidance for ALARA (as low as reasonably achievable) procedures. Despite the fact that cost levels exceeded the original estimate by 152.2 per cent, overall ALARA procedures reduced the originally estimated 1 007 man-rem cumulative integral dose to 155 man-rem over project life.

c) Reactor Pressure Vessel (RPV) Safety Analysis Report for Packaging (SARP)

The production of RPV SARP added $2.6 million to the project.

d) Decontamination, remove structures and engineering

These three elements reflect decisions to consolidate work to achieve contractor efficiency. The Decommissioning Plan was prepared to maximize competitive bidding by subcontractors. There were increases over the estimated costs for these three elements, but the increases were offset by contractor efficiencies.

2. Gentilly-1 decommissioning programme

Description of the project

Gentilly-1 (G-1), a 250 MWe heavy water moderated boiling light water cooled prototype reactor (CANDU-BLW)[1], is located on the banks of St. Lawrence river in the province of Quebec, Canada. G-1 was constructed starting in 1966, was put in service in 1972 and operated intermittently until 1978. In 1980 the station was placed in a state of preservation which could have permitted a restart. At this time, major retrofitting work was demanded by the Canadian regulatory authority, Atomic Energy Control Board (AECB), to bring the plant into line with current requirements. In 1982 a joint committee of Hydro-Québec (the provincial utility and the operator of G-1) and AECL (Atomic Energy of Canada Limited) recommended decommissioning against the rehabilitation of the plant on economic ground.

---

1. The boiling light water concept was developed as a variant of the standard CANDU-PHWR, having a higher potential efficiency and requiring a smaller heavy water inventory. However, because of the success of the CANDU-PHWR and the substantial increase in heavy water production in Canada, work on the CANDU-BLW cycle was not taken beyond the G-1.

Parts of 1982 and 1983 were devoted to engineering and economic studies involving decommissioning of G-1. A wide range of decommissioning scenarios was considered in the study. They varied from safe storage of the plant and disposal of all radiological hazards inside the plant with surveillance (Stage-1), to prompt dismantling of the plant and disposal of all equipment and structures for ultimate release of the site (Stage-3). After careful consideration, a decision was made to decommission the plant to a "static state" which is a variant of Stage-1 and delayed Stage-3 after a dormancy period (52.5 years is assumed). The plant will be decommissioned so as to satisfy the "unrestricted site use" (USU) criterion, which is described later, and will not be returned to a "greenfield" condition.

The work on a two year programme to bring G-1 to a static state began in March 1984. At the end of the programme, the site licence was amended and it became the Gentilly-1 Waste Storage Facility. The scope of the decommissioning programme included draining and drying of all systems, isolation and sealing of systems, cutting and capping of all penetrations to the reactor building. In the programme, access to the reactor building has been permanently shut off except for one airlock that can be made operable for periodic inspection. The spent fuel bay has been decontaminated, and a concrete slab poured on top of it. What was the bay, is now a non-contaminated utility room. All radioactive materials and components inside the service building have been removed, and the area was decontaminated and released to Hydro-Québec for a simulator training center. Although sometimes not considered a part of decommissioning, this project also included removal of 3 213 spent fuel bundles from the spent fuel bay and interim on site dry storage in concrete canister.

Decommissioning cost estimate

Total decommissioning costs of G-1 were estimated in 1983 at $186.82 million (Canada, 1983 dollars). The costs cover the activities from the reactor shutdown in 1979 to the accomplishment of Stage-3. Table 2.6 is a summary of the cost estimate. This estimate includes 12.4 per cent of contingency allowance.

Total labour requirements for the whole decommissioning project are estimated 761 man-years (approximately 1 522 000 man-hours). Direct hands-on labour's man-year are approximately 63 per cent of the total. Project management and Engineering man-years are approximately 37 per cent. Occupational dose limit is assumed 50 mSv per year to atomic radiation workers, and 5 mSv per year to other workers.

Radioactive waste removed will be 4 200 tonnes before any treatment and will be 10 940 cubic meters after packaging for disposal. The criterion for material release as non-radioactive is assumed as: 1) 5 mR/hr at 1 cm from the material, 2) no detectable loose contamination and 3) no fixed contamination detectable. Criteria for unrestricted site use (USU) are assumed as: 1) ALARA (as low as reasonably achievable), 2) no detectable loose contamination and 3) fixed contamination less than 0.00005 $\mu Ci/cm^2$ for alpha radioactive material and less than 0.0001 $\mu Ci/cm^2$ for beta and gamma radioactive material.

No large scale waste disposal facilities to accept decommissioning wastes are currently available in Canada. Low level waste generated during the decommissioning programme from 1984 to 1986, is stored in designated areas of the turbine building. Other active waste is stored in the reactor containment building, which is sealed. However, in the cost estimate, Chalk River Laboratory is assumed as a tentative disposal site. Transportation distance of radioactive waste is assumed as 1 000 kilometers. Unit costs for radioactive waste treatment & packaging are assumed as $370 (Canada, 1983 dollars) per cubic meters of packaged waste to be disposed; radioactive waste transportation costs, $88 (Canada, 1983 dollars) per cubic meter; radioactive waste disposal costs, $880 (Canada, 1983 dollars) per cubic meter. It should be remembered that this disposal cost is not based on a specific disposal site.

3.   KKN Niederaichbach

KKN Niederaichbach was a 100 MWe gas-cooled heavy water moderated reactor, which has a vertical pressure tube type core and no cooling tower. The reactor was constructed starting in 1966 at Niederaichbach, Germany, and operated from 1972 to 1974 by Kernforschungszentrum Karlsruhe GmbH. A "Safe enclosure" licence was issued in 1981. An application for a licence for total dismantlement (to a greenfield) had been filed in 1980 and the permission was granted at the end of June 1987. All dismantling operations will be completed in early 1994, then the site will be released for unrestricted use.

The turbine and condenser have been removed from the plant and are now operating at a coal-fired station. The turbine house of the plant is being used as a buffer storage facility for decommissioning waste. The dismantling of all contaminated components has been accomplished in 1990 and approximately 1 000 tonnes of carbon and stainless steel has been removed. The remote dismantling of the reactor core started in 1990 and is planned to be completed by early 1992. The removal of the inner layer of the activated concrete of the bioshield will take place in 1992.

Decommissioning cost estimate

The total decommissioning cost was estimated in 1989. The cost estimate was DM 200 million (1989 Marks). The costs may include research and development costs of related decommissioning techniques and do not include the cost for spent fuel reprocessing nor contingency allowances.

Although the plant operated only for 2 years (18 effective full power days), the contamination level in the primary coolant system is not significantly lower than in stations that have operated much longer. Therefore, the short operation period does not decrease the decommissioning cost substantially.

Total amount of decommissioning radioactive waste is assumed to be approximately 1 200 tonnes before any treatment and 1 200 cubic meters when packaged for disposal. The radioactive waste will be disposed of into a deep geological repository which will be built in the future. The assumed transportation distance of the waste is approximately 700 km (oneway). It is considered that the waste, whose radioactivity is less than 3.7 million Bq/tonne, and whose surface radioactivity is less than 0.37 Bq/cm^2 for beta & gamma radioactivity and less than 0.037 Bq/cm^2 for alpha radioactivity, is free for radiolodical control. Approximately 1 000 tonnes of contaminated material has been recycled for nuclear use.

Total labour requirements were estimated to be approximately 800 man-years. Average staff size during decommissioning was 70. The annual occupational exposure to decommissioning workers was estimated 0.6 man-Sv/year.

4.   Chinon-2

Description of the project

Chinon A2, a 250 MWe gas cooled graphite moderated reactor (GCR), is located in the vicinity of Avoine on the banks of Loire river in France. Chinon A2 was constructed starting in 1959 and was put into service in 1964, and operated until 1985. The total net electricity production was 23.6 billion KWh. The plant owner/operator is Electricité de France (EDF). Its moderator is 1 700 tonnes of graphite and the fuel is 250 tonnes of natural uranium. The reactor core is contained in a spherical steel vessel, 18 meters diameter, and is connected to 4 heat exchangers by means of 4 primary loops.

The decommissioning scenario chosen is to bring the plant to a "reinforced Stage-1" (henceforth, "level 1R"), followed by a reference dormancy period of at least 50 years. It is required to:

- isolate and confine the reactor block pile (reactor building) as well as those of the 4 exchangers;

- dismount the primary circuits between the reactor core and exchangers;

- dismount all the auxiliary circuits;

- have in-situ intermediate storage of dismounted radioactive materials;

- clean all the buildings other than reactor and exchanger buildings, by removing all materials stored therein (decontamination will not be systematically used).

The decommissioning project up to the level 1R started in 1986 and is still in progress. Dismantling of one of the primary loops is taking place now. The project is planned to be complete in 1992.

For technical/economical reasons, interim storage was chosen rather than final storage on an ANDRA station. In the matter of packaging, containers are made from shells of the primary circuit, which were cut into straight sections and were sealed off top and bottom with solid plates welded in place. Each container is loaded with scrap iron (approximately 4 tonnes) and is characterized. All contaminated and activated materials removed will be concentrated in the reactor building and the four exchanger buildings at the final state of decommissioning to the level 1R. The reactor building and the exchangers are kept under negative pressure. Reinforcement of the reactor block containment will enable the postponement of its dismantling.

Decommissioning cost estimate and backgrounds

The Chinon A2 dismantling operations have been conducted on a local scale with the means available on the site and therefore, did not have the benefit of any generic effect. The development and the use of techniques and procedures, and the strategic decisions in the decommissioning project have been realized in the special circumstances. Therefore, it should be noted that the induced costs can not be directly applied for generic cases.

The total cost of the decommissioning project up to the level 1R is estimated to be FF 230 million (1990 French Francs). By the end of December 1990, approximately 80 per cent of this total cost has been spent. Table 2.7 is a summary of the expenditure by December 1990. The cost estimate (and the realized expenditure) does not include costs for:

- shut-down operation;
- studies other than those led by the plant staff from 1986;
- defuelling and fuel back-end;
- administration and direction;
- exploitation taxes;
- purchase of re-usable tools;
- treatment of radioactive metallic wastes (stored provisionally on site).

Contingency allowance is not considered in the cost estimate.

Total labour requirements for direct hands-on labour by the end of December 1990, are around 335 000 man-hours (210 000 man-hours from EDF workers and 125 000 man-hours from external workers). Total labour rate is assumed as FF 120-500 per hour. An average EDF staff size during decommissioning will be 20 persons. 50 mSv/year is considered as the occupational dose limit.

In the case of the complete dismantlement of Chinon A2, it is estimated that total radioactive decommissioning waste will be 13 000 tonnes (before any treatment) and non-radioactive waste will be 250 000 tonnes. In the present decommissioning project to the level 1R, 1 200 tonnes of radioactive material, which correspond to waste from primary and annex circuits, will be treated and be put in intermediate storage. Although no free release criteria are defined at national level, special criteria have been authorized for the Chinon A2 project: 3.7 Bq/cm^2 and 1 Bq/g (these are only for non-activated materials). The unit cost of in-situ intermediate storage is estimated as approximately FF 5 000/m^3 using primary circuit pieces as containers.

The post-operational and containment work, which were carried out from 1986 to the end of 1990, resulted in 172 man-rem (1.72 man-Sv) exposure to the workers. Taking the remaining work into account, a total occupational exposure during the whole decommissioning work up to the level 1R will be 200 man-rem (200 man-Sv). For example, the dismantling work of a primary loop gives an average exposure of 12 man-rem (0.12 man-Sv) for an average intervention time of 12 000 hours.

Table 2.1 Cost/cost estimates for ongoing/terminated decommissioning projects of nuclear power plants

PROJECT	FACILITY DESCRIPTION					MODES OF DECOMMISSIONING	ESTIMATED DECOMMISSIONING COSTS		YEAR OF ESTIMATE	CPI(c) ESCALATOR APPLIED	EXCHANGE RATE OF JAN. 1990 (NCUs per USD)
	Type	Capacity (a) (MWe)	Start of construction	Start of operation	End of operation		As given in the response to the questionnaire (b)	in millions of US $ of Jan. 1990			
Shippingport (USA)	PWR	72(d)	1955	1957	1982	Stage-1 + 3 years + Stage-3	98.3 M USD [91.3 M USD(e) (1989)]	103.3	1989	1.051	-
Gentilly-1	CANDU-BLW	250	1966	1972	1978	Stage-1 + 52.5yrs + Stage-3	186.8 M CAD (1983)	216.4	1983	1.358	1.172
KKN-1	HWGCR	100	1962	1972	1974	Stage-3	200 M DM (1989)	121.4	1989	1.027	1.692
Chinon 2	GCR	250	1959	1964	1985	Stage-1(f)	230 M FF (1990)	39.9	1990	1	5.761

(a) Design gross capacity.
(b) The number in parenthesis is the base year for the money value.
(c) Calculated as x/y   x = the national consumer price index (CPI) for January 1990.
           y = the national CPI for January of the base year of the money value.
(d) Average.
(e) Real costs.
(f) "Reinforced stage-1" which will be followed by a reference dormancy period of at least 50 years.

Table 2.2 **Packaged waste shipment data from 1984 to 1989**

CONTENTS	VOLUME (ft3)	WEIGHT (lb.)	RADIOACTIVITY (Ci)
Asbestos	37 843	304 690	2.492
Compressed Waste	858	27 365	0.039
Concrete	1 826	115 680	0.078
Lead	2 017	137 353	0.176
Metallic Waste	63 793	2 467 813	41.593
Soil	1 867	69 430	1.443
Solidified Sludge	5 790	436 660	4.296
Solid Waste	74 973	1 838 603	26.548
Spent Resins	3 570	124 405	40.816
One-Piece Components	11 560	1 003 600	24.269
RPV* Package	9 800	1 798 000	16 467.000
TOTAL	213 897	8 368 599	16 608.750

* RPV : Reactor Pressure Vessel.

Table 2.3 **Reactor pressure vessel lift, shipment and disposal costs summary**

WORK ITEM	COSTS*	TASK %	PROJECT %
Load internals	0.57	5.5	0.6
Install reactor pressure vessel head	0.05	0.5	0.1
Preparation for lift	1.62	15.6	1.6
Lift, loading and transportation	5.31	51.4	5.4
Burial	0.16	1.6	0.2
Safety Analysis Report for Packaging	2.63	25.4	2.7
TOTAL	10.34	100.0	10.6

* In millions of 1989 US dollars.

Table 2.4 Summarized Shippingport project cost data (a)

(1989 US dollars in thousands)

WBS ELEMENTS AND DESCRIPTIONS		Project Plan Cost Baseline	Target Cost Baseline	Final Project Cost
1.0	ENGINEERING	6 066	6 066	6 066
2.1	OPERATIONS PROJECT MANAGEMENT	8 423	10 209	10 557
2.1.2	HANFORD SERVICE (waste burial)	631	2 150	2 438
2.1.5	PRV SARP*	0	0	2 626
2.3.1.1.1	SITE MANAGEMENT & SERVICES	7 478	7 188	6 769
2.3.1.1.2	OPERATIONS SUPPORT & SERVICES	22 072	23 298	23 522
2.3.1.1.3	ENGINEERING	854	1 297	1 130
2.3.1.1.4	PROCUREMENT	562	385	671
2.3.1.1.5	SOLID WASTE MANAGEMENT	2 377	1 954	2 145
2.3.1.1.6	SYSTEM OPERATIONS SUPPORT	1 667	1 742	1 745
2.3.1.1.7	UTILITIES	2 031	1 808	1 596
2.3.1.1.8	LIQUID WASTE MANAGEMENT	508	1 292	1 281
2.3.1.1	SITE MANAGEMENT & SUPPORT	37 549	38 964	38 859
2.3.1.2.2	SITE MOD. AND SERVICES	4 779	5 511	5 180
2.3.1.2.3	RPVI/NST** PREPARATION/LIFTING/TRANSPORT	5 655	6 478	6 468
2.3.1.2.4	PIPING & EQUIPMENT	6 795	6 582	6 576
2.3.1.2.5	REMOVE PRIMARY COMPONENTS	1 306	1 299	1 318
2.3.1.2.6	POWER & CONTROL SYSTEMS	575	607	532
2.3.1.2.7	STRUCTURES	4 930	5 886	5 929
2.3.1.2.8	CONTAINMENT CHAMBERS	1 355	335	367
2.3.1.2.9	DECONTAMINATION	1 322	2 100	2 274
2.3.1.2	DECOMMISSIONING ACTIVITIES	26 717	28 798	28 644
2.3.2	HOME OFFICE SUPPORT	2 574	1 738	1 617
	MANAGEMENT RESERVE	1 335	3 791	0
	DOC FEE	4 225	4 323	5 408
2.3	DECOMMISSIONING OPS. CONTRACTOR TARGET	72 400	77 614	74 528
	SHIPPINGPORT PROJECT TOTAL	98 300	98 300	91 349

* SARP   Reactor Pressure Vessel Safety Analysis Report for Packaging.
** RPVI/NST   Reactor Pressure Vessel and Neutron Shield Tank.

(a) This table is a summary of the estimated and actual costs of the Shippingport project. Only major project cost elements are summarized by Work Breakdown Structure (WBS) element. Accumulated project costs are not intended to be obtained by direct addition of each preceding line entry in the columns.

Table 2.5   **WBS* elements with high percent of change between TEC** and final cost**

(1989 US dollars in thousands)

WBS ELEMENTS AND DESCRIPTIONS		Project Plan Cost Baseline	Final Project Cost	Dollar Variations	Percent of Change
2.1.2	Hanford Burial Service	631	2 438	1 807	286.4
2.3.1.1.8	Liquid Waste Management	508	1 281	773	152.2
2.1.5	RPV SARP***	0	2 626	2 626	100.0
2.3.1.2.9	Decontamination	1 322	2 274	952	72.0
2.3.1.1.3	Engineering	854	1 130	276	32.3
2.3.1.2.7	Remove Structure	4 930	5 929	999	20.5

*   Work Breakdown Structure.
**   Total Estimated Cost developed from project planning.
*** Reactor Pressure Vessel Safety Analysis Report for Packaging.

Table 2.6   **Summarized decommissioning cost estimate of Gentilly-1**

(1983 Million Canadian dollars)

ITEM	COST ESTIMATE
Operation & Maintenance cost for 1979-1983	42.24
Spent Fuel Transfer to Intermediate "On site" spent fuel storage canisters	2.26
Station shutdown activities	10.90
Bringing G-1 to "static state conditions"	23.53
Dormancy of 52.2 years	26.5
Final Decommissioning to Stage-3 USU* estimate	81.39
TOTAL	186.82

*   USU : Unrestricted Site Use condition.

Table 2.7   Realized costs up to February 1991 for Chinon A2 dismantling project to "Reinforced Stage-1" (Provisional/15 February 1991)

COST ITEMS	MILLION FRANCS (1990 Value)	%
Reactor/primary circuit (studies, dismounting, decontamination, confinement)	17	9.4
Heat exchangers (studies, dismounting, decontamination, confinement)	6	3.3
Annex circuits (studies, dismounting, decontamination, confinement)	23	12.8
Electricity/instrumentation (studies, modifications)	3	1.7
Fuel manipulators (studies, dismounting, decontamination, confinement)	7	3.9
Overheads, waste management	56	31.1
Security, surveillance	10	5.6
General equipments, materials and tools	22	12.2
Maintenance of the site	15	8.3
Radioprotection	8	4.4
Fields and buildings	13	7.2
TOTAL	180	100

FUNDING SCHEME FOR DECOMMISSIONING IN MEMBER COUNTRIES

The information presented below comes primarily from answers to a questionnaire distributed to the Expert Group Members. The questionnaire asked for information such as an outline of the funding scheme, its target amount, collected amount and tax exemption/reduction system for the fund.

BELGIUM

Each electricity producing company is required to establish an internal fund for nuclear plant decommissioning. Funds are raised through annual contributions during plant lifetime (conventionally assumed as 20 years). Together with the interest accrued these contributions must, in 30 years from plant startup, make 12 percent of the investment cost (excluding interest during construction) presently needed to build such a plant. The interest calculation is based on rates customarily used for present worth calculations. Annual contributions to the fund are taken into account in the kWh prices for electricity.

CANADA

The Atomic Energy Control Board in its Regulatory Policy Statement (R-90) requires that all nuclear facility owner/operators develop their decommissioning plans with all associated actions assured by adequate financial planning. This statement does not determine any specific funding method and each facility owner/operator can choose its optimal method.

One method of decommissioning funding used by Canadian utilities is as follows:

The future cost of decommissioning is accounted for by means of equal annual charges to customers over the operating lifetime of the nuclear unit which, together with accumulated interest, would fund future expenditures as they are incurred.

FINLAND

According to the Nuclear Energy Act the power companies are responsible for the costs of their nuclear waste management (including decommissioning costs). The funds needed in future waste management investments must be collected into the State nuclear waste management fund (FUND).

Imatran Voima Oy (IVO) and Teollisuuden Voima Oy (TVO) must each prepare annually an estimate of the future costs for the management of all the wastes produced by the end of that year. The costs are estimated in real terms at the prevailing price level without any discounting. On the basis of these estimates the Ministry of Trade and Industry confirms the assessed liability of the company.

The fund target refers to the amount which the power companies must have in the FUND at a certain year. After n years of reactor operation the fund target is defined as follows:

$$FT = AL \times E_n/E_{25}$$

or $FT = AL \times n/25$

whichever gives higher figure.

FT = the fund target
AL = the assessed liability
n = number of years of operation
E_n = energy produced during n years of operation
E_{25} = energy to be produced during 25 years of operation with 75 per cent load factor.

In the first years of power production the fund target is much less than the assessed liability. The company must provide the State with securities to fill the gap between the assessed liability and fund target.

When the company invests in nuclear waste management, their assessed liabilty and fund target decrease. If the fund target decreases below the fund holding (the company's share of the fund: the money it has previously paid + interests), the difference is refunded to the company.

The assessed liabilities of 1989 and the fund targets of 1990 of the Finnish power companies are shown below:

	Assessed liability (MFIM)	Fund target (MFIM)
TVO	4 017	1 752
IVO	1 026	499

The big difference between IVO and TVO is based on different arrangements in spent fuel management. The fund so far collected covers the whole nuclear waste management and the share of decommissioning is not separately defined.

The fund contribution paid annually to the State nuclear waste management fund can be exempted from taxes as expenses in the annual income statement of the company.

FRANCE

Electricité de France (EDF) makes a monthly provision in the accounts, calculated per production unit, with the objective of including in the thermal production cost the cost chargeable to future decommissioning, spread over the various lifetimes, depending on the installation type:

- 20 years for gas graphite series power stations,
- 30 years for PWR type power stations.

For a given installation, the dismantling cost expressed in current french francs is determined by applying the following formula:

| Dismantling cost (in francs) | = | Reference cost per unit plant capacity (FF/kW) | x | Plant capacity (kW) | x | Monetary correction |

No matter what a reactor type/design is, the reference cost selected corresponds roughly to 15 per cent of the construction cost of a 1300 MW-PWR unit.

In the case of the Commissariat à l'Energie Atomique (CEA), its annual budget includes provisions for decommissioning facilities that have been shut down.

GERMANY

There is no official funding scheme in Germany. Utilities are responsible for all decommissioning actions including cost. They make provisions for liabilities and charges on a voluntary basis, based on estimated cost and facility operation time.

ITALY

No official funding system has been set up in Italy to set aside funds for decommissioning because the two national organisations which own Italian nuclear installations are state owned. The two organisations are ENEL (National Electricity Board) and ENEA (National R&D Board for Nuclear and other Alternative Energies).

JAPAN

Utilities started to reserve funds for decommissioning cost in 1989. A tax-free system for decommissioning funds, which was introduced in 1990, allows the utilities to reserve 85 per cent of total decommissioning cost with free of tax.

According to this system, a total decommissioning cost is identified based on the estimated weight of dismantled wastes of each plant. The total cost is calculated by the weight linear approximation based on the model plant's cost estimation results. The target amount of the fund is 85 per cent of the total decommissioning cost.

The annual contribution to the fund is calculated by multiplying target amount for reserve and the ratio of annually generated electricity to the total electricity which will be generated during 27 years operation with availability factor of 70 per cent.

SPAIN

An assessment on future costs of the back-end of the nuclear fuel cycle (including decommissioning of NPP) is made by ENRESA. These estimates are annually revised and presented in the General Radioactive Waste Plan (GRWP) which is subsequently submitted for Government approval.

In order to finance the costs to be incurred years later, a system of payment on account is established. The type of approach, for funding back-end activities is to collect a certain percentage (named quota) upon all electricity sales. Thus, incomes resulting from the quota, when added to the accumulating, interest fund, are just sufficient to cover the projected future expenditures.

The methodology used in calculating the quota is based on the principle that, incomes of one year are proportional to the electricity generated by NPP in this year. Figure 3.1 shows the steps followed to calculate the quota.

The Government, in the Royal-Decree authorising annually the electricity rates, establishes the value of the quota, according to the estimation made in the GRWP.

The target amount of the fund is US$800 M (1990) which is calculated with a discount rate of 3.5 per cent. As of now, 21 per cent of the total has been collected.

There is no tax exemption/reduction system for decommissioning fund in Spain.

SWEDEN

According to the Swedish legislation the costs for the back-end of the nuclear fuel cycle including the decommissioning of the nuclear power plants shall be borne by the reactor owners. To cover the future costs a fee is levied on the nuclear electricity production. The collected fees are paid to the state and collected in funds at the National Bank of Sweden. One fund is set up for each reactor owner. The funds are administered by The National Board for Spent Nuclear Fuel, SKN.

The fee shall be paid as long as the reactors are in operation and should be spread out equally over the total production period. The level of the fee is determined annually by the Government and is set separately for each reactor station. The decision by the Government is based on a proposal by SKN.

In order to provide a basis for the fee proposal the reactor owners are required, by law, to prepare annually a cost calculation for all the activities to take care of the spent fuel and the radioactive waste and to decommission the reactors and take care of the waste from decommissioning. This cost calculation is prepared by the Swedish Nuclear Fuel and Waste Management Company, SKB, and is presented in a report, the Plan report (ref), to SKN in June each year.

SKN then calculates the fee in such a way that it will have the same nominal value over the entire electricity production, that is it will ideally rise with the inflation. In making the proposal for the fee, SKN has to consider all relevant factors such as total costs, expected operation time of the reactors and interest on the money collected in the funds. As the costs are presented in the value of the calculation date the real rate of interest, that is the inflation corrected interest should be used.

In the 1990 cost calculation the total costs for the back-end are estimated to be 53 GSEK. Of this 11 GSEK is attributable to decommissioning and managing of decommissioning waste. The total fee for 1990 has on the average been 0.019 SEK/kWh. Of this about 0.004 SEK/kWh corresponds to decommissioning.

The total fund content at the end of 1990 is 7.6 GSEK. The money is used for the different waste management activities already under way, such as interim storage, transport and research and development.

UNITED KINGDOM

United Kingdom has no official funding scheme for decommissioning costs. Plant owner/operators are responsible for decommissioning costs.

Nuclear Electric

Since 1978/79 Nuclear Electric, and its predecessor the CEGB, has made specific financial provision in its annual accounts for decommissioning its nuclear power stations. These provisions are reviewed annually taking into account inflation and the latest decommissioning cost estimates. During the operating life of a nuclear power station sufficient provisions are made such that, at cessation of generation, there should be sufficient money available to cover all stages of decommissioning including the passive phase between Stage-2 and 3.

Nuclear Electric is currently in negotiation with the Inland Revenue (the UK's tax authority) concerning the treatment of nuclear provisions for tax purposes. The most likely outcome of these negotiations is that provisions will be treated as deductible for tax purposes in the year in which provision is made.

BNFL

Provisioning for decommissioning is shared between BNFL and its customers according to an agreed formula. BNFL's provisions are reinvested in the Company and BNFL has been able to demonstrate a sufficient rate of return including the anticipated treatment for taxation purposes to justify the assumed discount rate.

UNITED STATES

The U.S. Nuclear Regulatory Commission (NRC) requires nuclear power plant operators to establish a fund for decommissioning. The requirement is set out in US Title 10, Code of Federal Regulations, Part 50 (Section 50.33). Essentially the requirement is to accumulate in a segregated, and external fund a minimum amount in a range of US$75 to US$135 million in 1986 dollars. This is based on a specific formula for plant size and age plus an escalation adjustment. The intention in requiring a decommissioning fund is to ensure that adequate funding for decontamination will be available at the end of a plant's useful life and in this way to prevent potential risk to public health and safety. No costs specific to structural demolition or site restoration activities are required for inclusion in the fund amounts because NRC requirements pertain only to operator license termination matters. The utilities with existing licenses are required to implement funding provisions by July 1990. The total amount of the fund is to be available at commencement of a power plant's decommissioning. The fund source is current utility rate payers' payments. This approach is based on the principle that the consumers who benefit from electric power production should bear the costs of decommissioning the power plant at the end of its useful life.

Generally, US tax laws allow a utility to deduct from taxable income the dollar amounts put into a nuclear facilities decommissioning fund.

Figure 3.1. **CALCULATION OF THE QUOTA FOR YEAR N**

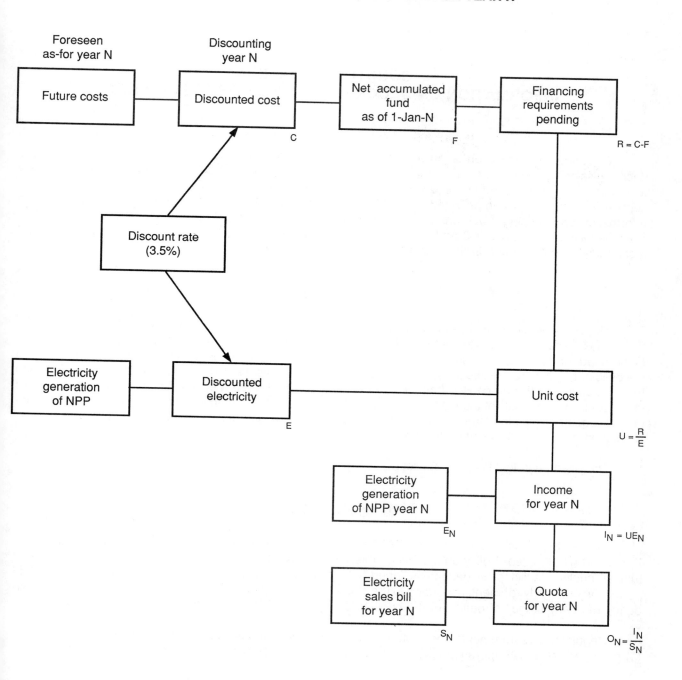

Annex 4

COUNTRY UP-DATE
INFORMATION ON DECOMMISSIONING COST ESTIMATES

This annex contains information on national decommissioning policies, decommissioning strategies and decommissioning regulations. The description focuses on the topics which are not described in the main text and which are important for decommissioning cost estimates in each country. It is intended to supplement the information provided in the 1986 NEA report which should be seen for further information. Information on financing provisions for decommissioning and on unrestricted release levels of radioactive waste is included in Annexes 3 and 4. It should be noted that "development and demonstration of related technology" is not included in the scope of this annex.

BELGIUM

The Eurochemic reprocessing facility, at the Mol-Dessel site, was shutdown in 1975. As from 1987, the site is used as the centralized interim storage for all conditioned radioactive wastes, and is also gradually being used for centralized waste processing. For this purpose, the site is being re-modelled and many of the buildings that were erected for specific purposes have to be dismantled. The project started in 1989 with the dismantling of the air cooling batteries. At the beginning of 1990, decommissioning has been started in the contaminated cells of the reprocessing plant themselves. In order to demonstrate the feasibility of the general decommissioning strategy, two buildings have been emptied and decontaminated as a pilot project. A detailed description is included in Annex 1.

The exemption level/criteria applicable to decommissioning wastes is not generally defined.

CANADA

The Atomic Energy Control Board (AECB), a Federal Government agency, has issued a Regulator Document (R-90) titled "Policy on the Decommissioning of Nuclear Facilities" in 1988. The AECB requires that all nuclear facilities be decommissioned satisfactorily in the interest of health, safety, security and protection of the environment, according to the plans approved by the AECB. Such plans shall be developed during the early stages of design of the nuclear facility and refined during the operating life of the facility.

Regulatory requirements have been specified for various stages of a facility's lifetime as (1) pre-operational, (2) operational, (3) post-operational, (4) post-decommissioning.

The AECB recognizes that the decommissioning programmes appropriate to specific nuclear facilities may vary greatly with facility type and, to a lesser extent, amongst facilities of similar type decommissioning of individual facilities may be accomplished in continuous programs or over discrete progressive or intermittent phases. It may thus include periods of "storage with surveillance".

FRANCE

Decommissioning operations in France are not subject to separate regulatory provisions, but constitute a special subset of regulations covering all nuclear facilities. No mandatory decommissioning timetable is stipulated, and no technical options are imposed: they are determined under the operator's responsibility. However, the following strategies are generally adopted corresponding to the category of the facility:

— total immediate decommissioning for fuel cycle plants, hot cells and pool reactors;

— partial immediate decommissioning for reactors which may be maintained in this condition over long time periods without incurring excessive costs, and which are designed to ensure containment of the residual radioactivity within a compact volume.

It is important to emphasize that these are general recommendations.

Even if an operator has decided to defer decommissioning of his facility, the decision is not irreversible. The decision may be reconsidered at any time in the light of changing political or economic contexts.

Radioactive waste management

Radioactive waste management in France depends on the radionuclide half-life, the activity level and the nature of the emitted radiation. The following classification is currently applicable:

— Short half-life (Category A) wastes

- Half-life <30 years: low or intermediate level wastes.
  Specific activity thresholds depending on the radionuclides present in the package are mandatory for storage.

- Half-life >30 years: very low-level wastes.
  α activity <37 GBq.t^{-1} for condition wastes,
  α activity <3.7 GBq.t^{-1} for entire storage facility.

- Disposal requirements:
  Category A wastes are placed in surface storage at a site authorized for a specified total radiological capacity.

— Long half-life wastes

- α-bearing (Category B) wastes.
  Wastes exceeding the authorized specific activity threshold for surface storage. Category B wastes are conditioned and placed in interim storage pending transfer to a final geological repository.

- Vitrified (Category C) wastes.
  These are primarily β,γ wastes, but also include very high α activity. They are cooled in interim storage prior to ultimate disposal in a geological repository.

- Disposal requirements:
  A research programme destined to qualify underground repository sites is now in progress under the responsibility of the French national agency for radioactive waste management (ANDRA).

Exemption level/criteria applicable to decommissioning wastes

French legislation has not yet fixed a level below which a body is not considered to be radioactive and no longer subject to regulatory control.

Legislation for radioactive components only applies as of 74 Bq/g or 100 Bq/g (for β,γ emitters) but these levels are not applied to unrestricted release.

Instructions and regulations applicable to power stations rest on the following principle that "all materials or material not having been subjected to neutron irradiation and not having been in contact with a contaminating fluid are considered as non-radioactive".

On a case by case basis, special items have been treated after being given authorization by the Central Protection Service against Ionizing Radiation (SCPRI).

FINLAND

The exemption level of 10 kBq/kg defined in the Finnish regulations is primarily meant for the exemption of certain types of operational wastes and is not directly applicable to decommissioning wastes. In a recent statement the Finnish Centre for Radiation and Nuclear Safety calls for further studies to define a proper level for the decommissioning wastes.

GERMANY

No exemption level for decommissioning waste is defined in german regulations. A level is defined in the radiation protection ordinance, but it is not adopted by the licensing authority. The exemption level in regulations is likely to be much lower than the level in the ordinance.

The Konrad Mine (crystalline rock) repository, in which decommissioning and reactor wastes will be disposed of, is expected to be ready in 1994.

JAPAN

Electric power companies started to reserve tax funds for decommissioning costs from the final settlement of accounts in FY-1989, and the tax-free system for the fund was established in FY-1990 (see Annex 3).

SPAIN

Low and intermediate level waste (LLW and ILW) generated by the application of radioisotopes in research, medicine, industry and agriculture as well as those generated by the operation of several nuclear power plants are stored in the ENRESA's (the National Radwaste Management Company's) facility of El Cabril, at present the only storage facility operating in Spain. This facility is being expanded in order to store all LLW and ILW regardless of its origin. The concept chosen is shallow burial in concrete structures providing sufficient capacity of waste produced up to the end of the year 2000.

Criteria and plans are being developed for decommissioning nuclear power plants and research reactors, as well as for stabilizing of tailings from mining and milling activities. For LWRs, total dismantling with 5 to 10 years is being contemplated for purposes of planning and cost estimating. Beginning of decommissioning activities would be 5 years after the final shutdown of each reactor. Specific decommissioning plans will be prepared closer to the shutdown time in a case by case basis. For Vandellos I (GCR), a different decommissioning strategy is considered (see Chapter 4).

An effort at regulatory level to establish an exemption level is ongoing. Concerning other criteria for unrestricted release, the applicable regulations provide "a priori" special situations to be exempted from the general licensing procedure. These regulations support authority's actions to exclude from regulatory contrils the releases of some very low radioactive materials on a case-by-case basis.

UNITED KINGDOM

Decommissioning strategy

Nuclear Electric (NE) has inherited from its successor, the Central Electricity Generating Board (CEGB), a reference case decommissioning strategy for its gas cooled reactors; Stage-1, then Stage-2 followed by Stage-3 after a storage period of 90 years. The cost estimates in the main text have been based on this strategy, against which the NE's decommissioning liability is assessed.

Recently the NE has had a team undertaking a major strategy review to establish if there are equally acceptable strategies for achieving the objectives of decommissioning at considerably lower cost. These studies have been completed and a recommendation has been made to senior management that a new strategy should be adopted. The recommended strategy is:

1. Stage-1 — Immediate fuel removal (typical duration years 0-5 from end of generation)
2. Stage-2 — Phase 1 • Preparation for care and maintenance (typically years 5-7 from end of generation)
    — Phase 2 • Care and maintenance (typically years 7-35 from end of generation)
    — Phase 3 • Safe-store structures construction (typically years 35-37 from end of generation)
3. Safestore care and maintenance
4. Stage-3 — Dismantling to greenfield status or in-situ decommissioning of the active plant and building typically years 135-145 from end of generation)

Provisions would be made against the more expensive dismantled Stage-3. However, it is intended to progress the lower cost in-situ decommissioning safety case to make it an available option for future generations. The proposed strategy results in considerably reduced discounted costs (approximately 60 per cent of the previous figure for an average Magnox station).

This proposed strategy is not yet approved by the NE, neither have accurate liabilities been calculated for the NE's 13 gas-cooled reactor stations, nor has regulatory approval been obtained at this stage. For the PWR currently under construction (Sizewell B), there is no present intention to change the policy.

Regulation

The Radioactive Substances (Substances of Low Activity) Exemption Order 1986 Statutory Instruments No. 1002 defines exemption limits relevant to decommissioning to a maximum specific activity of 0.4 Bq per gram. The possibility of relaxing this limit to 1 Bq per gram for wastes left behind on a decommissioned site if significant economic savings can be made is contained in "Radioactive Substances Act 1960: A Guide to the Administration of the Act 1982".

UNITED STATES

Waste management

Disposal of low level waste is by shallow land burial performed under provisions of the Low Level Radioactive Waste Policy Act of 1980 as amended in 1985. The intent of the act is the development by state regional compacts of low level waste disposal capacity. The vision at the time of enactment was a shared responsibility among the concerned state governments that would lead to new disposal facilities.

Since the act's amendment in 1985, the states have assumed substantially new roles. The states have written and enacted new legislation, thus creating interstate compacts and low level waste boards. Responsibility for low level waste disposal activities has been dispersed. In turn, this has led to regional disposal compact agreements, proposals for disposal facilities, and new state regulations for waste management.

The amendment to the act, notably the requirements for less waste and a surcharge to make inefficient disposal practice more costly, have motivated users to reduce wastes. However, though significant advances have been made, future disposal demand continues to exceed present disposal capacity. This is a serious concern but future demand capacity is a manageable situation, because the true demand for decommissioning waste disposal capacity will not occur until large scale physical decommissionings are performed sometime in the future. During the interim, there is sufficient time to plan and to develop appropriate capacity and methods for disposal of low level wastes.

Decommissioning rules

The U.S. Nuclear Regulatory Commission rules stipulate the technical and financial criteria for decommissioning licensed nuclear facilities. The decommissioning rule promulgates the following: the definition of decommissioning, acceptable decommissioning alternatives, planning for decommissioning, assurance of the availability of funds for decommissioning, and environmental review requirements related to decommissioning.

A three step process is required for the formation of a decommissioning plan and decommissioning cost estimates. First is the compilation of a decommissioning report to accompany a license application that indicates how reasonable assurance will be provided that funds will be available to decommission the facility (for already licensed facilities this was to have been completed by July 1990). This decommissioning report is to be updated on an annual basis.

Second is a preliminary decommissioning plan that is completed at or about five years prior to the projected end of plant operation. The preliminary decommissioning plan contains an up to date cost estimate and an assessment of major technical factors affecting the plan.

Third is the detailed decommissioning plan and cost estimate submitted within two years following permanent cessation of operations (and in no case less than one year prior to license expiration). This planning is required to precede regulatory approval for decommissioning and license termination.

Regulation

In the United States de minimis level is not defined in regulations. There is a waste classification system based on concentration of contamination per cubic meter of waste. The classification designation is arrived at by calculation based on the radionuclide content of the waste material and whether it is short or long lived.

Concerning decontamination, the radioactivity level to be achieved in decommissioning, as required by the U.S. Nuclear Regulatory Commission (NRC), is to residual levels that permit release of a facility for unrestricted use and for termination of the operator's license. These levels currently are set out in the U.S. Nuclear Regulatory Commission Regulatory Guide 1.86, Table I. Prior to release of facilities and materials for unrestricted use, the licensee is required to perform a comprehensive radiation survey to confirm that residual contamination is within acceptable levels.

The U.S. NRC requires the facility licensee to adhere to these levels; however, the U.S. Environmental Protection Agency (EPA) is responsible ultimately for setting limits for residual radiation that may remain. At this time, the EPA is developing standards for radiation protection criteria for the unrestricted release of nuclear facilities.

As a practical matter, the NRC in July 1990 disseminated a Below Regulatory Concern (BRC) policy. As a result of this policy, future decontamination limits may limit exposure to individual members of the public to 10 milli-rem per year. The policy establishes an individual dose criteria of 1 and 10 milli-rem per year (0.01 and 0.1 milli-Sievert per year).

The individual dose criteria of 1 milli-rem (0.01 milli-Sievert) per year applies to practices involving common consumer-use products that contain very small amounts of radioactivity (for example, house smoke detectors, calibration sources, and similar items). The individual dose criteria of 10 milli-rem (0.1 milli-Sievert per year) applies to defined exempted practices (for example, disposal of very low level radioactive wastes and other such well-defined practices). The collective dose (that is, the sum of individual of totals effective dose equivalents) criteria is 1 000 person-rem per year (10 person-Sievert per year). At this level of collective dose, the number of hypothetical health effects calculated for an exempted practice on an annual basis would be less than one.

The EPA uses 10^{-6} lifetime risk of cancer as the quantitative criterion of insignificance in sensitivity-of-measure, risk based guidelines for collective dose matters. The NRC did not develop the Below Regulatory Concern policy using this factor. The policy was developed based on an annual risk coefficient of 5×10^{-4} health effects per rem (5×10^{-2} per Sievert). Using this conservative annual risk coefficient factor, the 10^{-6} lifetime risk value translates to an annual dose of 0.01 to 0.1 milli-rem (0.0001 to 0.001 milli-Sievert) per year that an individual would incur from a continuous lifetime dose rate. The NRC believes that inclusion of individual doses below 0.1 milli-rem (0.001 milli-Sievert) per year introduces unnecessary complexity into collective dose assessments and could impute an unrealistic sense of significance and certainty for such dose levels.

While, this Below Regulatory Concern policy is a departure from current practice, this action is consistent with current regulatory philosophy to examine the need for a general policy on the appropriate criteria for release of radioactive materials from regulatory control. However, the licensee must apply for a license amendment for this.

The purpose in such considerations is the establishment of a framework within which the NRC will formulate rules or make licensing decisions to exempt from some or all regulatory controls certain practices involving small quantities of radioactive materials. The exemptions may involve the release of licensee controlled radioactive material either to the generally accessible environment or to persons who would be exempt from the Commission's regulation. For example, practices for which exemptions may be granted include, but are not limited to:

1. the release for unrestricted public use of land and structures containing residual radioactivity;
2. the distribution of consumer products containing small amounts of radioactive material;
3. the disposal of very-low radioactive waste at other than licensed disposal sites; and
4. the recycling of slightly contaminated equipment and materials.

This policy is expected to exempt specific practices from regulatory control. Particularly, in the situation where application or continuation of regulatory control is not necessary to protect the public health and safety and environment, and is not cost effective in future reducing risk.

COMMISSION OF THE EUROPEAN COMMUNITIES

The European Community is conducting R&D activities in the field of decommissioning of nuclear installations. The activities began with the 1979-1983 programme on the decommissioning of nuclear power plants which was followed by the 1984-88 programme on the decommissioning of nuclear installations. Presently, the 1989-93 programme on the decommissioning of nuclear installations is being implemented.

The 1989-93 programme has been started with the preparation of the four pilot projects, which concern:

- the Windscale Advanced Gas-Cooled Reactor;
- the KRB-A Boiling Water Reactor (Gundremmingen);
- the BR-3 Pressurized Water Reactor (Mol);
- the AT-1 fuel reprocessing facility (La Hague).

On the initiative of the Commission, the Group of Experts, set up under the terms of Article 31 of the Euratom Treaty, recommended clearance levels for the recycling of materials from the dismantling of nuclear installations. The recommended levels are directly applicable to steel scrap and equipment from nuclear power plants, but the methods by which they were derived can be applied for the development of criteria for other valuable metals, such as copper and aluminium, and other nuclear installations. The recommended clearance levels are:

For beta/gamma radiation:

- 1 Bq/g averaged over a maximum mass of 1 000 Kg; to avoid the inclusion of highly active items within the average mass, there is an additional requirement that no single item may exceed 10 Bq/g.

- 0.4 Bq/cm^2 for non-fixed contamination on accessible surface, averaged over 300 cm^2 or over the relevant surface area if it is less than 300 cm^2; for fixed contamination the mass activity concentration clearance level of 1 Bq/g is assumed to apply.

For alpha radiation:

- 0.04 Bq/cm^2 measured over any area of 300 cm^2 of any part of the surface.

For both beta/gamma and alpha concentrations, if doubt exists for non-accessible surfaces, the activity must be assumed to be higher than the respective clearance level (see Radiation Protection Series No. 43, CEC-DG XI/3134/88).

Annex 5

INTERNATIONAL PROGRAMME FOR THE EXCHANGE OF SCIENTIFIC AND TECHNICAL INFORMATION CONCERNING NUCLEAR INSTALLATION DECOMMISSIONING PROJECTS

Introduction

In the light of growing international concern for decommissioning of nuclear facilities, the Nuclear Energy Agency (NEA) of the OECD has been carrying out various activities in this field from the 1970s. The general atmosphere of co-operation developed among Member countries by the large number of contacts and exchanges carried out within the NEA framework created the favourable climate for the gathering of a consensus around a proposal to set up under the NEA a broad co-operative programme for the exchange of technology between important decommissioning projects in Member countries. This led to the setting up of the Co-operative Programme for the Exchange of Scientific and Technical Information Concerning Nuclear Installation Decommissioning Projects.

The programme started in 1985 with 10 decommissioning projects from 7 countries. Since then, the programme completed very successfully its first 5-year term and all the current participants (17 decommissioning projects from 8 countries) unanimously agreed to extend the duration of the programme for another 5 year term. Three other projects were included in the programme, leading to the current total number of 20 participating projects. A review report of the first term of five years is being prepared.

Scope of the programme

The basic scope of the Co-operative Programme involves an exchange of scientific and technical information between decommissioning projects. This includes project descriptions and plans, data obtained from research and development associated with the participating projects, data resulting from the execution of projects plans, and lessons learned from such execution. Such exchange does also include technical visits to the participating projects.

Institutional framework of the programme

During preparatory discussion for setting up the programme, it was agreed that this kind of co-operation would be best achieved within the framework of a formal agreement sponsored by such an international organisation as the NEA. This agreement was negotiated between 1984 and 1985, and entered into force on 18th September 1985, for a duration of five years. The success achieved by the programme in this first term led the participants to agree to renewal of the agreement for a second term of five years.

The participation in the programme is relatively open; in fact, it is not limited to participants operating or planning important decommissioning projects, but institutions having a specific interest and possibly future plans in decommissioning are also admitted as observers. There is, however, a difference in the level and degree of access to the information exchange. The observers have only access to a first level of exchanges including plans, strategies, rationale for the choice of approaches and methods as well as information of a general nature, whilst a second level, including detailed exchanges of specific data and other more in-depth forms of co-operation, is reserved to the full participants. The participation of decommissioning projects and/or observers from non-OECD countries can be considered and must be agreed upon by the governing body of the programme and authorised by the OECD management.

For the implementation of the Co-operative Programme, the agreement provided for the setting up of a governing body and a technical body. The governing body, called Liaison Committee (LC), is comprised of representatives of all participants, including the observers, and is responsible for the general conduct of the programme, including direction and supervision of the work programme, establishment of criteria for dissemination of the information exchanged or generated within the programme, approval of changes in membership, etc. The Secretariat of the LC is provided directly by the NEA Secretariat.

Although technical discussions are held in the LC, especially on topics of generic interest, the bulk of the actual exchange of scientific and technical information concerning the participating projects is carried out within the Technical Advisory Group (TAG), which is composed of technical managers and other senior specialists from the participating projects.

Projects in the Co-operative Programme

A list of the participating projects together with some basic information is presented in the following table. As the table shows, the programme covers 13 reactors, six fuel reprocessing plants and one isotope facility.

Table 5.1 **Projects in the Co-operative Programme**

	FACILITY	TYPE	POWER OR THRU-PUT	PROJECT TIME SCALE	DECOMMIS-SIONING OPTION*
1	Eurochemic Reprocessing Plant, Dessel, Belgium	Reprocessing of fuel	180 t (1966-1974)	1989-2004	Stage-3
2	BR-3, Mol, Belgium	PWR	41 MWt		Stage-3 (partial)
3	Gentilly-1, Canada	Heavy water moderated boiling light water cooled prototype reactor	250 MWe	1984-1986	Variant of Stage-1
4	NPD, Canada	PHWR CANDU prototype	25 MWe	1987-1988	Variant of Stage-1
5	Tunney's Pasture Facility, Canada	Radioisotope research and handling facility	N/A	1990-2005	Stage-3
6	Rapsodie, Cadarache, France	Experimental sodium cooled fast reactor	20 MWt	1992	Stage-2
7	G2, Marcoule, France	GCR, Electricity and nuclear materials production	250 MWt	1993	Stage-2
8	AT1, La Hague, France	Pilot reprocessing plant for FBR	2 kg/day	1981-1992	Stage-3
9	Kernkraftwerk Niederaichbach (KKN), Landshut, Germany	Gas-cooled heavy water moderated reactor	100 MWe	1992	Stage-3
10	MZFR, Karlsruhe, Germany	PHWR	50 MWe	2001	Stage-3
11	Kernkraftwerk Lingen (KWL), Germany	BWR (with super-heater)	520 MWt	1985-1988	Stage-1
12	Garigliano, Italy	BWR (dual-cycle)	160 MWe	1985	Stage-1 for main containment
13	Japan Power Demonstration Reactor (JPDR), Tokai, Japan	BWR	90 MWt	1986-1993	Stage-3
14	JAERI, Experimental reprocessing facility, Tokai, Japan	Experimental reprocessing facility		1990-1998	
15	Windscale Advanced Gas Cooled Reactor, Sellafield, UK	AGR	100 MWt	1983-1998	Stage-3
16	BNFL, Co-precipitation Plant, Sellafield, UK	Production of mixed Pu and UO_2 fuel		1986-1990	Stage-3
17	B204 Primary Separation Plant, Sellafield, UK	Fuel reprocessing facility	Metal 500 t/y Oxide 140 t/y	1990-2005	Stage-2
18	Shippingport, United States	PWR	72 MWe	1985-1989	Stage-3
19	West Valley Demonstration LWR Fuel Project, United States	Reprocessing plant for LWR fuel (waste vitrification and removal phase)	640 t (1966-1972)	1996-2004	Stage-3
20	Experimental Boiling Water Reactor (EBWR), United States	BWR	100 MWt	1986-1995	Stage-3

* The decommissioning options are defined according to the IAEA classification (see IAEA Technical Report Series No.230).

Annex 6

SELECTED WAGES STATISTICS FOR OECD COUNTRIES

This annex contains selected wages statistics for OECD countries by the International Labour Organisation (ILO). The numbers indicated are for "earnings", which include direct wages and salaries, remuneration for time not worked (excluding severance and termination pay), bonuses and gratuities and housing and family allowances paid by the employer directly to this employee.

The concept of earnings, as applied in ILO statistics, relates to remuneration in cash and in kind paid to employees, as a rule at regular intervals, for time worked or work done together with remuneration for time not worked, such as for annual vacation, other paid leave or holidays. Earnings excludes employer's contribution in respect of their employees paid to social security and pension schemes and also the benefits received by employees under these schemes. Earnings also exclude severance and termination pay.

Statistics of earnings relate to employees' gross remuneration, i.e. the total before any deductions are made by the employer in respect of taxes, contributions of employees to social security and pension schemes, life insurance premiums, union dues and other obligations of employees.

In the following tables data is given for construction and manufacturing industries of selected OECD countries. Original data are quoted from ILO Year Book on Labour Statistics, 1989-90 (ILO, 1990).

Table 6.1 Earnings in construction industries

COUNTRY	IN NATIONAL CURRENCIES (A)		IN 1990 JANUARY US DOLLARS* (B)	
BELGIUM	299.02	BF/hr (1988)(a)	8.79	USD/hr
CANADA	16.10	CAD/hr (1989)(b)	14.10	USD/hr
FINLAND	48.74	FIM/hr (1988)	13.50	USD/hr
FRANCE	39.96	FF/hr (1987)(c)	7.44	USD/hr
GERMANY	18.89	DM/hr (1989)(d)	11.34	USD/hr
ITALY	9 544	Lire/hr (1985)(e)	9.70	USD/hr
JAPAN	373 211	Yen/month (1989)(f)	16.23	USD/hr
SPAIN	561	Pesetas/hr (1987)(g)	5.95	USD/hr
SWEDEN	93.01	SEK/hr (1989)(h)	15.75	USD/hr
UNITED KINGDOM	4.867	£/hr (1989)(i)	8.16	USD/hr
UNITED STATES	13.37	USD/hr (1989)	14.39	USD/hr

*   (B) = (A) x (CPI inflator) / (NCU per US dollar in January 1990).
     1 working-month is assumed to be 160 working-hours.

(a)   Wage earners, October 1988.
(b)   Employees paid by the hour.
(c)   October 1987.
(d)   Males, including family allowances paid directly by the employers.
(e)   Including the value of payments in kind.
(f)   Employees, including family allowances and mid- and end-of-year bonuses.
(g)   Employees.
(h)   Adults, second quarter of 1989, including holidays and sick-leave payments and the value of payments in kind.
(i)   Full-time workers on adult rates of pay, October 1989.

Table 6.2 Earnings in manufacturing industries

COUNTRY	IN NATIONAL CURRENCIES (A)		IN 1990 JANUARY US DOLLARS* (B)	
BELGIUM	309.78	BF/hr (1988)(a)	9.11	USD/hr
CANADA	13.54	CAD/hr (1989)(b)	11.85	USD/hr
FINLAND	43.45	FIM/hr (1989)(c)	11.29	USD/hr
FRANCE	40.97	FF/hr (1987)(d)	7.62	USD/hr
GERMANY	19.10	DM/hr (1989)(e)	11.46	USD/hr
ITALY	9 451	Lire/hr (1985)(f)	9.60	USD/hr
JAPAN	336 648	Yen/month (1989)(g)	14.64	USD/hr
SPAIN	678	Pesetas/hr (1987)(h)	7.19	USD/hr
SWEDEN	79.30	SEK/hr (1989)(i)	13.43	USD/hr
UNITED KINGDOM	4.916	£/hr (1989)(j)	8.24	USD/hr
UNITED STATES	10.47	USD/hr (1989)	10.76	USD/hr

*   (B) = (A) x (CPI inflator) / (NCU per US dollar in January 1990).
     1 working-month is assumed to be 160 working-hours.

(a)   Wage earners, October 1988.
(b)   Employees paid by the hour.
(c)   Including mining, quarrying and electricity.
(d)   October 1987.
(e)   Including family allowances paid directly by the employers.
(f)   Including the value of payments in kind.
(g)   Employees, including family allowances and mid- and end-of-year bonuses.
(h)   Employees.
(i)   Adults, second quarter of 1989, including holidays and sick-leave payments and the value of payments in kind.
(j)   Including quarrying, full-time workers on adult rates of pay, October 1989.

GLOSSARY

Activity: The activity, A, of an amount of radioactive nuclide in a particular energy state at a given time is the quotient of dN by dt, where dN is the expectation value of the number of spontaneous nuclear transformations from that energy state in the time interval dt:

$$A = \frac{dN}{dt}$$

The special name for the SI unit of activity is becquerel (Bq): $1 \text{ Bq} = 1 \text{ s}^{-1}$. The special unit of activity, curie (Ci), may be used temporarily: $1 \text{ Ci} = 3.7 \times 10^{10}$ Bq (exactly).

ALARA: As low as reasonably achievable, economic and social factors being taken into account. A basic principle of radiation protection taken from the Recommendations of the International Commission on Radiological Protection (ICRP), ICRP Publication No. 26.

Alpha-Bearing Waste: Waste containing one or more alpha-emitting radionuclides, usually actinides, in quantities above acceptable limits. The limits are established by the national regulatory body.

Becquerel (Bq): The SI unit of radioactivity, equivalent to 1 disintegration per second (approx. 2.7×10^{-11} Ci).

Biological Shield: Physical barriers to reduce exposure to living organisms, e.g. the concrete shield around the reactor.

Capacity (Electrical): The load for which a generating unit is rated, either by the user or by the manufacturer.

Containment: The retention of radioactive material in such a way that it is effectively prevented from becoming dispersed into the environment or only released at an acceptable rate.

Contamination, Radioactive: A radioactive substance in a material or place where it is undesirable. Surface contamination is the result of the deposition and attachment of radioactive materials to a surface.

Curie (Ci): A unit of activity equal to 3.7×10^{10} Bq.

Decontamination: The activities to remove or decrease the radioactivity of reactor components, shield walls and so on.
Chemical methods: Removing contaminants by washing or dissolving with acids, solvents or others.
Physical methods: Removing contaminated parts by paring with mechanical devices or others.

Discounting: A procedure to convert the value of money earned or spent in the future to a present value. If one had $A and it could be invested to earn interest at a real money rate "r" per annum, in "t" years it would increase to become $A(1+r)^t$. A sum of $B earned or spent in t years time can be said to have a present value of $B/(1+r)^t$. The "r" is entitled a "discount rate".

Disposal: The emplacement of waste materials in a repository, or at a given location, without the intention of retrieval. Disposal also covers direct discharge of both gaseous and liquid effluents into the environment.

Exposure: Any exposure of persons to ionising radiation. A distinction is made between:

    a) external exposure, being exposure to sources outside the body;
    b) internal exposure, being exposure to sources inside the body;
    c) total exposure, being the sum of the external and internal exposures.

Gigawatt (GW): One million kilowatts. GW(e) refers to electric power, whereas GW(th) refers to thermal power.

Greenfield Conditions: The conditions in which all facilities of the nuclear power plant are dismantled and removed from its site.

Half-Life, Radioactive: For a single radioactive decay process, the time required for the activity to decrease to half its value by that process. (After a period equal to ten half-lives, the activity has decreased to about 0.1 per cent of its original value).

High-Level Waste: The highly radioactive waste material that results from the reprocessing of spent nuclear fuel, including liquid waste produced directly in reprocessing and any solid waste derived from the liquid and which contains a combination of TRU waste and fission products in such concentration as to require long-term isolation.

Interim Storage (Storage): A storage operation for which:

    a) monitoring and human control are provided, and
    b) subsequent action involving treatment, transportation, and final disposition is expected.

Intermediate-Level Waste (or medium-level waste): Waste of a lower activity level and heat output than high-level waste, but which still requires shielding during handling and transportation.

Long-Lived Nuclide: For waste management purposes, a radioactive isotope with a half-life greater than about 30 years.

Low-Level Waste: Waste which, because of its radionuclide content, does not require shielding during normal handling and transportation. (See alpha-bearing waste and high-level waste for other possible limitations).

Megawatt (MW): One thousand kilowatts (see also Gigawatt).

Nuclear Fuel: Fissionable and/or fertile material for use as fuel in a nuclear reactor. Fissionable materials can be fissioned by neutrons. Fertile material can be converted to fissionable material by absorbing neutrons.

Nuclear Power Plant: A single- or multi-unit facility in which heat produced in a reactor(s) by the fissioning of nuclear fuel is used to drive a steam turbine(s) which in turn drives an electric generator.

Nuclear Reactor: An apparatus in which the nuclear fission chain can be initiated, maintained, and controlled so that energy is released at a specific rate. The reactor apparatus includes fissionable material (fuel) such as uranium or plutonium; fertile material; moderating material (unless it is a fast reactor); a containment vessel; shielding to protect personnel; provision for heat removal; and control elements and instrumentation.

Plant: The physical complex or buildings and equipment, including the site.

Radioactive Decay: A physical phenomenon that each radioactive material decease its radioactivity at a rate during a unit period. Due to this phenomenon, each radioactive material half its radioactivity in a specified period, which is entitle a "half-life". The half-life of Co-60, which is the dominant source of the radioactivity in the reactor vessel and its internals, is about 5.3 years.

Radioactive Material: A material of which one or more constituents exhibit radioactivity.

Radioactive Waste: Any material that contains or is contaminated with radionuclides at concentrations or radioactivity levels greater than the exempt quantities established by the authorities and for which no use is foreseen.

Radioactivity: The property of certain nuclides of spontaneously emitting particles or gamma radiation, or of emitting X-radiation following orbital electron capture, or of undergoing spontaneous fission.

Reactor Internals: Structural components which are placed inside a reactor vessel. Core supports, coolant flow distribution components and neutron reflector components are included. Irradiated fuels are excluded. These internals are highly activated due to the neutron irradiation from the core and their radioactivity is the highest among the waste from decommissioning.

Reactor Types:

AGR (Advanced Gas-Cooled Reactor): A gas-cooled reactor with stainless steel-clad slightly enriched uranium dioxide fuel elements; cooled by carbon dioxide.

BWR (Boiling-Water Reactor): A light-water reactor in which water, used as both coolant and moderator, is allowed to boil in the core. The resulting steam can be used directly to drive a turbine.

CANDU (Canadian Deuterium-Uranium reactor): A pressurized heavy water reactor of Canadian design, which uses natural uranium as a fuel and heavy water as a moderator and coolant.

GCR (Gas-Cooled Reactor): A reactor in which a gas such as air, carbon dioxide or helium is used as a coolant.

Magnox Reactor: A gas-cooled reactor cooled by carbon dioxide; using magnox alloy (a magnesium alloy with low aluminium content) as a cladding material for fuel.

HTGR (High Temperature Gas-Cooled Reactor): A graphite-moderated helium-cooled advanced reactor.

LWR (Light Water Reactor): A nuclear reactor that uses ordinary water as both a moderator and a coolant and utilises slightly enriched uranium-235 fuel. There are two commercial LWR types: the boiling water (BWR) and the pressurized water reactor (PWR).

PWR (Pressurized-Water Reactor): A light-water reactor in which heat is transferred from the core to a heat exchanger via water kept under high pressure, so that high temperatures can be maintained in the primary system without boiling the water. Steam is generated in a secondary circuit.

Reactor Vessel: (in Water Cooled Reactor such as PWR and BWR)

A tank of thick steel which contains a reactor core. Because this tank is irradiated directly by neutrons from the core, this tank has strong radioactivity when the plant is closed. Due to this intensive radioactivity and physical strength, dismantling is one of the most important activities in the decommission.

Regulatory Authority or Regulatory Body: An authority or system of authorities designated by the government of a Member State as having the legal authority for conducting the licensing process, for issuing of licenses and thereby for regulating the siting, design, construction, commissioning, operation, shutdown, decommissioning and subsequent control of nuclear facilities (e.g. waste repositories) or specific aspects thereof. This authority could be a body (existing or to be established) in the field of nuclear-related health and safety or mining safety or environmental protection, vested with such legal authority, or it could be the government or a department of the government.

Safe Storage: A period of time starting after the initial decommissioning activities of preparation for safe storage cease and in which surveillance and maintenance of the facility takes place. The duration of time can vary from a few years up to an order of 100 years. (See stage of decommissioning).

Shallow-Land Disposal (e.g. shallow-land burial): Disposal of radioactive waste, with or without engineered barriers, above or below the ground surface, where the final protective covering is of the order of up to about 10 meters thick. Some Member countries consider shallow-land disposal to be a mode of storage rather than a mode of disposal.

Stage of Decommissioning: The term stage implies a state or condition of a facility after decommissioning activities:

Stage-1 — storage with surveillance;
Stage-2 — restricted site release;
Stage-3 — unrestricted site release.

Surveillance: Includes all planned activities performed to ensure that the conditions at a nuclear installation remain within the prescribed limits; it should detect in a timely manner any unsafe condition and the degradation of structures, systems and components which could at a later time result in an unsafe condition. These activities may comprise:

a) monitoring of individual parameters and system status;
b) checks and calibrations of instrumentation;
c) testing and inspection of structures, systems and components;
d) evaluation of the results of items a) and c).

Transuranic (TRU) Waste: Waste containing quantities of nuclides above agreed limits having atomic numbers above 92. The limits are established by national regulatory bodies.

Treatment of Waste: Operations intended to benefit safety or economy by changing the characteristics of the waste. Three basic treatment concepts are:

a) volume reduction;
b) removal of radionuclides from the waste;
c) change of composition.

Wastes from Decommissioning Nuclear Power Plants: Spent fuels and radioactive wastes generated during the operating period are not included in the wastes. Radioactive wastes from decommissioning of nuclear power plants belong to the Intermediate Level Waste or Low Level Waste. High Level Waste is not produced in the process of decommissioning.

Waste Management: All activities, administrative and operational, that are involved in the handling, treatment, conditioning, transportation, storage and disposal of waste.

Annex 8

LIST OF EXPERT GROUP MEMBERS

Belgium	Mr. M. DUSONG	Tractebel Energy Engineering
	Mr. L. LEYS	NIRAS/ONDRAF
	Mr. L. TEUNCKENS	Belgoprocess N.V.
Canada	Mr. G. PRATAPAGIRI	Atomic Energy of Canada Ltd.
Finland	Mr. H. HÄRKÖNEN	Imatran Voima Oy
	Dr. J. VIRA　(Chairman)	Teollisuuden Voima Oy
France	Mr. G. BETSCH	EDF-CPN St. Laurent
	Mr. L. CHAUDON	Commissariat à l'Energie Atomique (Marcoule)
	Mr. G. COSTE	CEN (Saclay)
	Mr. J.-P. GATINEAU	EDF SEPTEN
	Mr. F. MUNOZ	EDF/SPT
	Mr. J. ROGER	Commissariat à l'Energie Atomique (Marcoule)
Germany	Dr. U. LÖSCHHORN	Kernforschungszentrum Karlsruhe
Italy	Mr. M. CONTI	ENEA
	Mr. T. VITIELLO	ENEL
Japan	Mr. H. FUJIHARA	Tokyo Electric Power Company
	Mr. K. FUJIKI	Japan Atomic Energy Research Institute
	Mr. S. MURAKAMI	Tokyo Electric Power Company
	Mr. Y. OGATA	Mitsubishi Heavy Industries Ltd.
	Mr. Y. SOTOME	Hitachi Ltd.
	Mr. M. TSUTAGAWA	Toshiba Corporation
	Mr. S. YANAGIHARA	Japan Atomic Energy Research Institute
Spain	Mr. ABREU	ENRESA
	Mr. R.G. GARCIA-SUELTO	ENRESA

Sweden	Mr. H. FORSSTRÖM	Swedish Nuclear Fuel & Waste Management Company
	Mr. S. PETTERSSON	Swedish Nuclear Fuel & Waste Management Company
	Mr. N. RYDELL	National Board for Spent Nuclear Fuel
United Kingdom	Mr. S. GORDELIER	Nuclear Electric plc
	Mr. J. HOLDER	Ex British Nuclear Fuels plc
	Mr. P. McNICHOLAS	UKAEA Technology
	Mr. A. SHEIL	British Nuclear Fuels plc
	Mr. J. SHEPHERD	UKAEA Technology
	Mr. G. WREN	UKAEA Technology
United States	Mr. M.L. McKERNAN	Roy F. Weston, Inc.
	Mr. W.E. MURPHIE	U.S. Department of Energy
Commission of the European Communities	Mr. L.V. BRIL	Directorate General XVII, Nuclear Energy
	Mr. F. PFLUGRAD	Directorate General XII, Decommissioning Programme
IAEA	Mr. P.L. DE	Division of Nuclear Fuel Cycle and Waste Management
OECD Nuclear Energy Agency	Mr. G.H. STEVENS	Nuclear Development Division
	Mr. K. TODANI	Nuclear Development Division
	Mr. M. YASUI (Secretary)	Nuclear Development Division
NEA Co-operative Programme on Decommissioning	Mr. S. MENON	Studsvik Nuclear

WHERE TO OBTAIN OECD PUBLICATIONS – OÙ OBTENIR LES PUBLICATIONS DE L'OCDE

Argentina – Argentine
CARLOS HIRSCH S.R.L.
Galería Güemes, Florida 165, 4° Piso
1333 Buenos Aires   Tel. 30.7122, 331.1787 y 331.2391
Telegram: Hirsch-Baires
Telex: 21112 UAPE-AR. Ref. s/2901
Telefax:(1)331-1787

Australia – Australie
D.A. Book (Aust.) Pty. Ltd.
648 Whitehorse Road, P.O.B 163
Mitcham, Victoria 3132   Tel. (03)873.4411
Telefax: (03)873.5679

Austria – Autriche
OECD Publications and Information Centre
Schedestrasse 7
D-W 5300 Bonn 1 (Germany)   Tel. (49.228)21.60.45
Telefax: (49.228)26.11.04
Gerold & Co.
Graben 31
Wien I   Tel. (0222)533.50.14

Belgium – Belgique
Jean De Lannoy
Avenue du Roi 202
B-1060 Bruxelles   Tel. (02)538.51.69/538.08.41
Telex: 63220   Telefax: (02) 538.08.41

Canada
Renouf Publishing Company Ltd.
1294 Algoma Road
Ottawa, ON K1B 3W8   Tel. (613)741.4333
Telex: 053-4783   Telefax: (613)741.5439
Stores:
61 Sparks Street
Ottawa, ON K1P 5R1   Tel. (613)238.8985
211 Yonge Street
Toronto, ON M5B 1M4   Tel. (416)363.3171
Federal Publications
165 University Avenue
Toronto, ON M5H 3B8   Tel. (416)581.1552
Telefax: (416)581.1743
Les Publications Fédérales
1185 rue de l'Université
Montréal, PQ H3B 3A7   Tel.(514)954-1633
Les Éditions La Liberté Inc.
3020 Chemin Sainte-Foy
Sainte-Foy, PQ G1X 3V6   Tel. (418)658.3763
Telefax: (418)658.3763

Denmark – Danemark
Munksgaard Export and Subscription Service
35, Nørre Søgade, P.O. Box 2148
DK-1016 København K   Tel. (45 33)12.85.70
Telex: 19431 MUNKS DK   Telefax: (45 33)12.93.87

Finland – Finlande
Akateeminen Kirjakauppa
Keskuskatu 1, P.O. Box 128
00100 Helsinki   Tel. (358 0)12141
Telex: 125080   Telefax: (358 0)121.4441

France
OECD/OCDE
Mail Orders/Commandes par correspondance:
2, rue André-Pascal
75775 Paris Cédex 16   Tel. (33-1)45.24.82.00
Bookshop/Librairie:
33, rue Octave-Feuillet
75016 Paris   Tel. (33-1)45.24.81.67
                          (33-1)45.24.81.81
Telex: 620 160 OCDE
Telefax: (33-1)45.24.85.00 (33-1)45.24.81.76
Librairie de l'Université
12a, rue Nazareth
13100 Aix-en-Provence   Tel. 42.26.18.08
Telefax : 42.26.63.26

Germany – Allemagne
OECD Publications and Information Centre
Schedestrasse 7
D-W 5300 Bonn 1   Tel. (0228)21.60.45
Telefax: (0228)26.11.04

Greece – Grèce
Librairie Kauffmann
28 rue du Stade
105 64 Athens   Tel. 322.21.60
Telex: 218187 LIKA Gr

Hong Kong
Swindon Book Co. Ltd.
13 - 15 Lock Road
Kowloon, Hong Kong   Tel. 366.80.31
Telex: 50 441 SWIN HX   Telefax: 739.49.75

Iceland – Islande
Mál Mog Menning
Laugavegi 18, Pósthólf 392
121 Reykjavik   Tel. 15199/24240

India – Inde
Oxford Book and Stationery Co.
Scindia House
New Delhi 110001   Tel. 331.5896/5308
Telex: 31 61990 AM IN
Telefax: (11)332.5993
17 Park Street
Calcutta 700016   Tel. 240832

Indonesia – Indonésie
Pdii-Lipi
P.O. Box 269/JKSMG/88
Jakarta 12790   Tel. 583467
Telex: 62 875

Ireland – Irlande
TDC Publishers – Library Suppliers
12 North Frederick Street
Dublin 1   Tel. 744835/749677
Telex: 33530 TDCP EI   Telefax: 748416

Italy – Italie
Libreria Commissionaria Sansoni
Via Benedetto Fortini, 120/10
Casella Post. 552
50125 Firenze   Tel. (055)64.54.15
Telex: 570466   Telefax: (055)64.12.57
Via Bartolini 29
20155 Milano   Tel. 36.50.83
La diffusione delle pubblicazioni OCSE viene assicurata
dalle principali librerie ed anche da:
Editrice e Libreria Herder
Piazza Montecitorio 120
00186 Roma   Tel. 679.46.28
Telex: NATEL I 621427
Libreria Hoepli
Via Hoepli 5
20121 Milano   Tel. 86.54.46
Telex: 31.33.95   Telefax: (02)805.28.86
Libreria Scientifica
Dott. Lucio de Biasio 'Aeiou'
Via Meravigli 16
20123 Milano   Tel. 805.68.98
Telefax: 800175

Japan – Japon
OECD Publications and Information Centre
Landic Akasaka Building
2-3-4 Akasaka, Minato-ku
Tokyo 107   Tel. (81.3)3586.2016
Telefax: (81.3)3584.7929

Korea – Corée
Kyobo Book Centre Co. Ltd.
P.O. Box 1658, Kwang Hwa Moon
Seoul   Tel. (REP)730.78.91
Telefax: 735.0030

Malaysia/Singapore – Malaisie/Singapour
Co-operative Bookshop Ltd.
University of Malaya
P.O. Box 1127, Jalan Pantai Baru
59700 Kuala Lumpur
Malaysia   Tel. 756.5000/756.5425
Telefax: 757.3661
Information Publications Pte. Ltd.
Pei-Fu Industrial Building
24 New Industrial Road No. 02-06
Singapore 1953   Tel. 283.1786/283.1798
Telefax: 284.8875

Netherlands – Pays-Bas
SDU Uitgeverij
Christoffel Plantijnstraat 2
Postbus 20014
2500 EA's-Gravenhage   Tel. (070 3)78.99.11
Voor bestellingen: Tel. (070 3)78.98.80
Telex: 32486 stdru   Telefax: (070 3)47.63.51

New Zealand – Nouvelle-Zélande
GP Publications Ltd.
Customer Services
33 The Esplanade - P.O. Box 38-900
Petone, Wellington
Tel. (04)685-555   Telefax: (04)685-333

Norway – Norvège
Narvesen Info Center - NIC
Bertrand Narvesens vei 2
P.O. Box 6125 Etterstad
0602 Oslo 6   Tel. (02)57.33.00
Telex: 79668 NIC N   Telefax: (02)68.19.01

Pakistan
Mirza Book Agency
65 Shahrah Quaid-E-Azam
Lahore 3   Tel. 66839
Telex: 44886 UBL PK. Attn: MIRZA BK

Portugal
Livraria Portugal
Rua do Carmo 70-74, Apart. 2681
1117 Lisboa Codex   Tel.: 347.49.82/3/4/5
Telefax: (01) 347.02.64

Singapore/Malaysia – Singapour/Malaisie
See "Malaysia/Singapore" - Voir «Malaisie/Singapour»

Spain – Espagne
Mundi-Prensa Libros S.A.
Castelló 37, Apartado 1223
Madrid 28001   Tel. (91) 431.33.99
Telex: 49370 MPLI   Telefax: 575.39.98
Libreria Internacional AEDOS
Consejo de Ciento 391
08009 - Barcelona   Tel. (93) 301-86-15
                          Telefax: (93) 317-01-41
Llibreria de la Generalitat
Palau Moja, Rambla dels Estudis, 118
08002 - Barcelona   Telefax: (93) 412.18.54
Tel. (93) 318.80.12 (Subscripcions)
(93) 302.67.23 (Publicacions)

Sri Lanka
Centre for Policy Research
c/o Mercantile Credit Ltd.
55, Janadhipathi Mawatha
Colombo 1   Tel. 438471-9, 440346
Telex: 21138 VAVALEX CE   Telefax: 94.1.448900

Sweden – Suède
Fritzes Fackboksföretaget
Box 16356, Regeringsgatan 12
103 27 Stockholm   Tel. (08)23.89.00
Telex: 12387   Telefax: (08)20.50.21
Subscription Agency/Abonnements:
Wennergren-Williams AB
Nordenflychtsvägen 74, Box 30004
104 25 Stockholm   Tel. (08)13.67.00
Telex: 19937   Telefax: (08)618.62.32

Switzerland – Suisse
OECD Publications and Information Centre
Schedestrasse 7
D-W 5300 Bonn 1 (Germany) Tel. (49.228)21.60.45
Telefax: (49.228)26.11.04
Librairie Payot
6 rue Grenus
1211 Genève 11   Tel. (022)731.89.50
Telex: 28356
Subscription Agency – Service des Abonnements
Naville S.A.
7, rue Lévrier
1201 Genève   Tél.: (022) 732.24.00
Telefax: (022) 738.48.03
Maditec S.A.
Chemin des Palettes 4
1020 Renens/Lausanne   Tel. (021)635.08.65
Telefax: (021)635.07.80
United Nations Bookshop/Librairie des Nations-Unies
Palais des Nations
1211 Genève 10   Tel. (022)734.14.73
Telex: 412962   Telefax: (022)740.09.31

Taiwan – Formose
Good Faith Worldwide Int'l. Co. Ltd.
9th Floor, No. 118, Sec. 2
Chung Hsiao E. Road
Taipei   Tel. 391.7396/391.7397
Telefax: (02) 394.9176

Thailand – Thaïlande
Suksit Siam Co. Ltd.
1715 Rama IV Road, Samyan
Bangkok 5   Tel. 251.1630

Turkey – Turquie
Kültur Yayinlari Is-Türk Ltd. Sti.
Atatürk Bulvari No. 191/Kat. 21
Kavaklidere/Ankara   Tel. 25.07.60
Dolmabahce Cad. No. 29
Besiktas/Istanbul   Tel. 160.71.88
Telex: 43482B

United Kingdom – Royaume-Uni
HMSO
Gen. enquiries   Tel. (071) 873 0011
Postal orders only:
P.O. Box 276, London SW8 5DT
Personal Callers HMSO Bookshop
49 High Holborn, London WC1V 6HB
Telex: 297138   Telefax: 071 873 2000
Branches at: Belfast, Birmingham, Bristol, Edinburgh,
Manchester

United States – États-Unis
OECD Publications and Information Centre
2001 L Street N.W., Suite 700
Washington, D.C. 20036-4910 Tel. (202)785.6323
Telefax: (202)785.0350

Venezuela
Libreria del Este
Avda F. Miranda 52, Aptdo. 60337, Edificio Galipán
Caracas 106   Tel. 951.1705/951.2307/951.1297
Telegram: Libreste Caracas

Yugoslavia – Yougoslavie
Jugoslovenska Knjiga
Knez Mihajlova 2, P.O. Box 36
Beograd   Tel.: (011)621.992
Telex: 12466 jk bgd   Telefax: (011)625.970

Orders and inquiries from countries where Distributors
have not yet been appointed should be sent to: OECD
Publications Service, 2 rue André-Pascal, 75775 Paris
Cedex 16, France.

Les commandes provenant de pays où l'OCDE n'a pas
encore désigné de distributeur devraient être adressées à :
OCDE, Service des Publications, 2, rue André-Pascal,
75775 Paris Cédex 16, France.

75880–7/91

OECD PUBLICATIONS, 2 rue André-Pascal, 75775 PARIS CEDEX 16
PRINTED IN FRANCE
(66 91 09 1) ISBN 92-64-13552-9 - No. 45681 1991